A Companion to Justinian's *Institutes*

A Companion to Justinian's *Institutes*

Edited by
Ernest Metzger

Duckworth

First published in 1998 by
Gerald Duckworth & Co. Ltd.
61 Frith Street, London W1V 5TA
Tel: 0171 434 4242
Fax: 0171 434 4420
Email: duckworth-publishers.co.uk

ISBN 0 7156 2798 8 hbk
ISBN 0 7156 2830 5 pbk

A catalogue record for this book is available from the British Library.

Typeset by Ray Davies
Printed in Great Britain by
Redwood Books Ltd, Trowbridge

Contents

Table of Contents

3. PROPERTY

D.L. Carey Miller

4. SUCCESSION

William M. Gordon

5. OBLIGATIONS

Robin Evans-Jones
Geoffrey MacCormack

6. ACTIONS

Ernest Metzger

7. CRIMINAL TRIALS

O.F. Robinson

For David Daube

Preface

A textbook of law is usually able to speak for itself without help from anyone. Even after so many centuries, Justinian's *Institutes* can often still speak for itself. It has always been the best and clearest introduction to Roman law, and time and again it gives relief from the more difficult books that try to explain it. It remains the law student's first textbook, as Justinian promised. Of course Justinian could not have known that his law would endure in so much of the modern world. He therefore could not have known that later generations would take such an interest in the classical sources of law he relied on, the methods of the classical jurists, and the development of the law in Rome over a thousand years. The *Institutes* is therefore still the law student's first textbook, but the student now needs to know a great deal more than Justinian tells him.

This book is a companion to the translation of the *Institutes* prepared by Peter Birks and Grant McLeod, and published by Duckworth and Cornell in 1987. It addresses all the principal subjects discussed in the *Institutes* and gives a thorough description of the law relating to each subject. It is therefore in every respect a textbook of Roman law – a second textbook, after the *Institutes* itself. In presenting this book we note that the translators took enormous care both to produce a beautiful and accurate translation and to render the language as a whole as consistent as possible. Accordingly the present book hopes to be faithful not only in following Justinian's treatment, but also in matching the translators' care in presenting Justinian's technical vocabulary. To help the reader follow the vocabulary of the two languages, the Glossary and the English-Latin Word List associate Latin terms with their adopted English equivalents, nearly all of which follow those the translators have established. Passages from the *Institutes* quoted in this book and citations to ancient works also follow, of course, the Birks and McLeod edition.

Each chapter is divided into sections. Where a passage in the *Institutes* is the principal source for discussion in a section, a citation to that passage is given in the section heading. Where the principal source is a passage from Gaius' *Institutes*, a citation to Gaius is given instead. Each author has provided a Select Bibliography of works pertinent to his subject, to supplement the General Bibliography. Footnotes have been kept to a minimum, and cross-references within and among chapters are generous. Full tables of authorities are included at the end.

I am grateful to Jilly Arbuthnott, who did the initial editing, and to Lucy Metzger, who performed the first copy-edit on the manuscript. My thanks also go to Peter Birks and Robin Evans-Jones, who as my predecessors took great trouble to ensure that this book would be produced.

Ernest Metzger
University of Aberdeen

Abbreviations

AJ = *Acta Juridica.*
ANRW = H. Temporini and W. Haase (eds), *Aufstieg und Niedergang der römischen Welt* (Berlin, 1972 –).
BIDR = *Bullettino dell' Istituto di Diritto Romano.*
Birks (ed), *New Perspectives* = P. Birks (ed), *New Perspectives in the Roman Law of Property: Essays for Barry Nicholas* (Oxford, 1989).
Birks and McLeod (eds), *Institutes* = P. Birks and G. McLeod (eds), *Justinian's Institutes* (Duckworth and Cornell, 1987; repr. 1994).
Buckland, *Textbook* = W.W. Buckland, *A Textbook of Roman Law from Augustus to Justinian* 3rd ed. rev. P. Stein (Cambridge, 1963).
C. = *Codex of Justinian*
CLJ = *Cambridge Law Journal.*
CQ = *Classical Quarterly.*
D. = *Digest of Justinian.*
G. = *Institutes of Gaius.*
IJ = *The Irish Jurist.*
J. or *Institutes* = *Institutes of Justinian.*
JR = *Juridical Review.*
JRS = *Journal of Roman Studies.*
Kaser, *RPR* I, II = M. Kaser, *Das römische Privatrecht* 2nd ed. (Munich, 1971, 1975), 2 vols.
Lenel, *EP* = O. Lenel, *Das Edictum Perpetuum* 3rd ed. (Leipzig, 1927).
LQR = *Law Quarterly Review.*
MacCormick and Birks (eds), *The Legal Mind* = N. MacCormick and P. Birks (eds), *The Legal Mind: Essays for Tony Honoré* (Oxford, 1986).
Nicholas, *Introduction* = B. Nicholas, *An Introduction to Roman Law* (Oxford, 1962).
Nov. = *Novel.*
RHDFE = *Revue Historique du Droit Français et Étranger.*
RIDA = *Revue Internationale des Droits de l'Antiquité.*
Roman Statutes I, II = M.H. Crawford (ed), *Roman Statutes* (*Bulletin of the Institute of Classical Studies*, suppl. 64) (London, 1996), 2 vols.
SALJ = *South African Law Journal.*
SDHI = *Studia et Documenta Historiae et Iuris.*
Stein and Lewis (eds), *Studies in Memory of Thomas* = P. Stein and A.D.E. Lewis (eds), *Studies in Justinian's Institutes in memory of J.A.C. Thomas* (London, 1983).
Thomas, *Textbook* = J.A.C. Thomas, *Textbook of Roman Law* (Amsterdam, 1976).

Tul L Rev = *Tulane Law Review*.
TvR = *Tijdschrift voor Rechtsgeschiedenis*.
Ulpian, *Rules* = *Ulpiani liber singularis regularum* [J. Baviera and J. Furlani (eds), *Fontes Iuris Romani Antiqui* II 2nd ed. (Florence, 1940) pp. 262-301].
ZPE = *Zeitschrift für Papyrologie und Epigraphik*.
ZSS (rom. Abt.) = *Zeitschrift der Savigny-Stiftung für Rechtsgeschichte*, romanistische Abteilung.

Contributors

D.L. Carey Miller, Professor of Property Law, University of Aberdeen.

Robin Evans-Jones, Professor of Jurisprudence, University of Aberdeen.

William M. Gordon, Douglas Professor of Civil Law, University of Glasgow.

Geoffrey MacCormack, Emeritus Professor of Jurisprudence, University of Aberdeen.

Ernest Metzger, Lecturer in Jurisprudence, University of Aberdeen.

O.F. Robinson, Reader in Law, University of Glasgow.

1. Sources

Geoffrey MacCormack

I. Introduction

The first two titles of book 1 are concerned with certain preliminaries which loosely can be described as 'sources of law', although they describe distinct species of law and the general moral principles underlying the positive law, as well as identifying the particular mechanisms by which the positive law was constituted. The treatment is fuller and more theoretical than that found in the *Institutes* of Gaius and, as one would expect, owes a great deal to the writings of the classical jurists excerpted in the *Digest*, particularly those of Ulpian.

The treatment of these preliminaries goes from the general to the specific. First, we have what may be termed a statement of the principal moral objectives which law should strive to realise. The opening sentence of the first title focuses upon the importance of maintaining the individual rights of men, taken more or less verbatim from Ulpian (D.1.1.10 pr):

> Justice is an unswerving and perpetual determination to acknowledge all men's rights. (J.1.1 pr).

Read in a modern context, this resounding statement would be interpreted as an affirmation of the sanctity of human rights. In the context of the *Institutes* it must be given a very much more limited scope in view of the fact that among the rights secured by justice in Roman law was that which permitted one person to hold another as slave.

After the general definition of justice there follows a statement, again literally following Ulpian (D.1.1.10.1), of the main principles which a just legal system should seek to implement, in particular the duty not to harm others and the duty to render each his due. Such principles constitute the foundation of any system of natural law. They are found, for example, in the forefront of the expositions of St Thomas Aquinas and Hugo Grotius.

Paragraph 4 of title 1 makes a double distinction, each element of which has to be understood from a different perspective. A distinction is first drawn between public and private law. This is a classification of the whole of the law applicable in the Empire into two kinds of rules: some pertaining to the organisation of the state, others to the well-being of the individual. The content of public law is not further explained. In the passage in the *Digest* from which the distinction has been taken (D.1.1.1.2), Ulpian specifies that public law comprises religious affairs, the priesthood, and the offices of state.[1] A second distinction is drawn among the law of nature, the law of all peoples, and the law of the state. This distinction is made in relation to private law alone, and like the first distinction looks like a classification of kinds of law. But in fact there is an unstated shift of perspective since the law of nature and the law of all peoples, unlike the law of the state, are not just divisions or varieties of private law; they are also, and more importantly, sources from which the rules of private law are derived. We have to consider in what sense exactly they are sources.

At this point it is appropriate to introduce a distinction between sources in the sense of a body of material, such as morality or public opinion which has induced the legislators or the judges to introduce a new rule, and sources in the sense of the precise mechanisms and procedures through which valid rules of law enter the legal system, such as the procedures for the enactment of a statute.[2] The question which arises is, in which of these senses does each of the three kinds of private law count as a source of that law? The law of the state (*ius civile*) is not a source in either sense. It is that part of private law which the Roman state has specifically developed for the well-being of its citizens. On the other hand, the law of nature (*ius naturale*) and the law of all peoples (*ius gentium*) do seem to have the character at least of sources in the first sense. Each may perhaps be described as a reservoir of principles upon which the state may draw in establishing, through a relevant procedure, rules for its citizens. These two varieties of law cannot, however, be regarded as a source in the second sense, that is, regarded as the product of a procedure for the introduction of new rules into the system. The Roman legal system does not appear either to have specified criteria by which rules of nature or of all peoples were to be identified, or to have provided that such rules, when identified, were to form part of the system.

II. The Three Components of Private Law
(J.1.2 pr - 2)

In title 2 the three laws which comprise private law are more fully elucidated. A word of caution to the student coming to the *Institutes* for the first time is appropriate. Each of these varieties of law has a number of different senses, not all of which are specifically mentioned in the *Institutes*. In any given instance of one of these terms, its meaning has to be gathered from the particular context in which it is used.

(a) Natural law
(J.1.2 pr)

The opening paragraph of the title, following Ulpian (D.1.1.1.3), defines the law of nature as that which nature has taught all animals (instinct). Hence, the institution of marriage and the rearing of children can be credited to the law of nature. Justinian omits a further point made by Ulpian (D.1.1.1.4), namely that the distinction between the law of nature and the law of all peoples lay precisely in the fact that the former applied equally to humans and animals, the latter only to humans. We have to assume that, both for Ulpian and Justinian, natural law in the sense of instinct contributed to Roman marriage only the notion of a habitual coming together of one man and one woman for the purpose of procreating children, rather than contributing any of the rules which stipulated how a marriage was to be accomplished or which defined the status of children born to the wife. Instinct adverts to practices or habits, not to behaviour prescribed or regulated by rules.[3]

This is not the only definition of natural law enunciated in the *Institutes*. Two further references to natural law are made, each going well beyond the scope of the first definition. In paragraph 1, when distinguishing between the law of all peoples (*ius gentium*) and the law of the state (*ius civile*), Justinian, now following the account given by Gaius in his *Institutes* (G.1.1), states that 'the law which natural reason makes for all mankind is applied the same everywhere'. Here in effect we have an uncritical fusion between natural law and the law of all peoples. One can see how the fusion comes to be made. Mankind everywhere is endowed by nature with the power to reason and hence, through the exercise of that reason, should everywhere produce the same rules. What this approach overlooks is that man does not necessarily exercise his reason to the same degree and, further, that some of the rules and practices which are universally found cannot necessarily be ascribed to reason. Why should Justinian have switched from a version of natural law based on instinct to one based on reason? The answer is supplied by the context in which the second version of natural law is introduced. Justinian is now speaking of people who have laws and customs by which they are governed. Such laws and customs cannot be derived merely from instinct, since they imply the notion of a rule. What permits the establishment of rules by humans is the operation of reason. Hence Justinian like Gaius finds it appropriate to place the notion of reason in the forefront of his explanation.

Finally, almost at the end of the title in paragraph 11, Justinian offers a third definition of natural law. There is still the implication that it is indistinguishable from the law of all peoples: it is said to be 'observed uniformly by all peoples'. But two new features are now presented. Natural law is 'sanctioned by divine providence', and 'lasts forever', being immune from change. The second feature follows from the first. Because natural law is established by divine providence, it must necessarily be unchangeable. Humans cannot alter what God has established for their benefit. This definition, not found among the extant writings of the classical jurists,[4] reflects the fact that the Empire in Justinian's time was Christian. The context, again, explains why the *Institutes* at this particular point has recourse to the notion of rules. Consideration of the issue (which laws of the state can be changed?) prompts the observation that certain laws, in virtue of their divine origin, are unchangeable.

(b) The law of all peoples (J.1.2.1)

The meaning of the phrase *ius gentium* (the law of all peoples) is elucidated by way of contrast to that of *ius civile* (the law of the state). The linking of these two laws reflects an old historical connection, no longer relevant in the time of Justinian, which arguably possessed greater significance for Gaius from whose *Institutes* (G.1.1) this section of Justinian's account is drawn. This connection, mentioned explicitly neither by Gaius nor Justinian, is explained below. As it stands, the opening section of paragraph 1 simply divides the law of any people – strictly the reference should be only to private law – into two parts, that derived from the law of all peoples, here taken as synonymous with the law of nature, and that constituted by the rules which the state has made for itself, rules not necessarily shared by the peoples of other states.

At the end of paragraph 2 Justinian says that slavery and almost all forms of contract derive from the law of all peoples. Two points should be noted about the various examples he gives. The first is that slavery, as already noted by Ulpian in D.1.1.1.4, is specifically named as belonging to this law but not to the law of nature (see Chapter 2, section II (opening) below). Clearly the law of nature is here being used in a sense different from that of 'rules flowing from natural reason', since these have already been equated with the law of all peoples. One explanation is that the words 'by the law of nature all men were initially born free' advert to the initial definition of natural law as that which nature has instilled in all animals. Yet there is still something odd in ascribing the fact of being born free to the operation of an instinct. We may in fact have here an allusion to a fourth sense of natural law, best glossed as a 'state of nature' in which men were believed to have lived before there were states, governments, or laws.

The second point is the disparity in the kinds of examples adduced. Some (wars, captivity, slavery) refer to certain general practices or institutions which can plausibly be deemed to be of universal occurrence. But others (the contracts of sale, hire, partnership, deposit, and loan) appear, at first sight, to belong rather to the law which the Roman state has made for itself. It can hardly be contended that every state had the same rules governing these particular contracts as the Romans. Nevertheless, the inclusion of the particular contracts mentioned as examples of the law of all peoples can perhaps be defended on the ground that something resembling institutions of selling, hiring, forming partnerships, lending, and depositing can be found everywhere, even though there is considerable variation in detail.

Another factor may also explain the inclusion of contracts as examples of the law of all peoples. This is the technical and ancient sense of *ius gentium* in contrast to *ius civile* to which we have already referred. Originally, for example at the time of the *Twelve Tables* around 450 BC, the law of Rome (the *ius civile*) was wholly derived from statute or custom and applied only to Roman citizens. Later, as the state expanded and as citizens began to deal commercially with foreigners, it became necessary to devise rules that might suitably be applied to transactions involving foreigners. Little is known about the actual process, either before or after the creation of the office of the foreign praetor in 242 BC, by which rules were devised to facilitate transactions between foreigners and citizens. Some use may have been made of the rules and practices of those foreign peoples with whom the Romans had a commercial relationship, particularly the Greeks, but it seems that, on the whole, there was an extension to foreigners of rules already in use among citizens. Rules that were seen as generally adaptable, in virtue of their simplicity and lack of technical requirements, were applied to foreigners and later appeared in the edict of the foreign praetor. For example, the contracts of sale, hire, partnership, deposit, or loan had probably already evolved for use by citizens before their transfer to the field of transactions between foreigners and citizens.[5] The rules contained in the edict of the foreign praetor came to be regarded as 'the law of all peoples' in the sense that they were intrinsically capable of application to all peoples, not just the Romans. Thus, they might appropriately be cited as examples of the law of all peoples in its broader, more theoretical sense.

(c) The law of the state
(J.1.2.2)

The core sense of the phrase 'law of the state' (*ius civile*) was always that given in the *Institutes* of Gaius (G.1.1) and Justinian (J.1.2.1), namely, the law which each people had established for itself as peculiar to it. But the content of the notion saw an important expansion in the course of time. As we have already noted, the law peculiar to the Roman state was originally that derived from statute or custom, and applied only to Roman citizens. With the gradual expansion of the small city state and the increasing complexity of its legal system, the law applied in Rome came to be derived from a greater variety of sources, including the edicts of the foreign and urban praetors. Hence in one sense the rules introduced by the edict of the foreign praetor constituted just as much a part of the law peculiar to the state as those found in the urban edict. Technically the position remained that some legal institutions such as conveyance by mancipation (see Chapter 3, section III(b) below) or promising by *sponsio* (see Chapter 5, section III(a) below) were open only to Roman citizens. Such institutions were classified as belonging to 'the law of the state', while other institutions, such as conveyance by simple delivery (*traditio*, see Chapter 3, section III(d) below) and almost all the contracts (see Chapter 5, sections II-V below), were regarded as belonging to 'the law of all peoples' in the sense that they were open to non-citizens as well as to citizens. After the extension of the citizenship to all free inhabitants of the Empire by the constitution of the emperor Caracalla (AD 211-17) in AD 212 this technical distinction in effect disappeared.

A further contrast made by the classical jurists, though not one mentioned in the *Institutes*, is that between the law of the state and the edicts of the magistrates (see section III(d) below). This was a distinction based upon the derivation of the rules that constituted the law of the state in its broad sense. Some were derived from statute and custom (the law of the state in its narrow sense), others from the later source constituted by the edicts of the magistrates, which were classified as belonging to the magisterial law or *ius honorarium* in contrast to the *ius civile*. The latter term itself in this context came to be understood as comprising not only the rules derived from statute and custom, but also those contributed by juristic opinion, resolutions of the senate, or pronouncements of the emperors.[6]

III. Sources of Written Law

(J.1.2.3 - 8)

In title 2, paragraph 3 Justinian turns to an examination of the distinct processes by which rules of private law are established. His treatment is similar to that of Gaius, with the exception that he makes a preliminary distinction between written and unwritten law. The specific sources mentioned in this paragraph relate to written law only, unwritten law or custom being discussed briefly in paragraphs 9 and 11. The sources of written law which Justinian identifies correspond to those listed by Gaius (G.2.2), but differ somewhat from those mentioned by the late classical jurist Papinian whose account has been preserved in the *Digest* (D.1.1.7). We have already noted the contrast between the law of the state (*ius civile*) and the law of all peoples (*ius gentium*) in the technical sense of rules applied only to Roman citizens and rules applied also to foreigners. Papinian utilises a related, though still different, distinction in his classification of sources. He describes the law of the state as the law derived from acts, plebeian statutes, resolutions of the senate, imperial pronouncements, and the answers of the jurists. This is contrasted with the praetorian or magisterial law, that is, the rules introduced in the edicts of the magistrates, the object of which was to supplement and correct the civil law for the sake of public utility. Justinian, for whom the particular distinction between the civil and the magisterial law possessed no importance, chose rather to follow the broader treatment of Gaius which listed the edicts of the magistrates along with acts, plebeian statutes, resolutions of the senate, imperial pronouncements, and answers given by the jurists as sources of private law. We will examine each source in turn.

(a) Statutes
(J.1.2.4)

Acts and plebeian statutes (or 'plebiscites') as sources of law were important mainly in the period of the republic. Both were enactments of an assembly of the Roman people, but the assemblies which could enact an act differed in their composition from the assembly which could enact a plebeian statute.[7] An act might be enacted only by a formal assembly of the whole free male population comprising both patricians and plebeians. The two assemblies with effective law-making powers were the *comitia centuriata* and the *comitia tributa*, the former being the more important and the more often used for legislation. In both, the people assembled not as individuals but in groups. The *comitia centuriata* was composed of centuries, that is, military units differentiated according to wealth, and the *comitia tributa* of tribes, that is, units drawn from each geographic district. Each unit had one vote determined by a majority of its members. Women were excluded from participation, but sons within authority, if adult, were eligible. The important point to note is that neither assembly might initiate legislation. It might merely vote on a proposal put to it by a competent magistrate, its function being to affirm or reject without the power to modify. Only the highest magistrates, consuls or praetors, could propose legislation to the *comitia centuriata*, but in addition the curule aediles could propose to the *comitia tributa*.

There also existed another assembly of the whole people called the *comitia curiata* in which the people (the men) were organised into groups (*curiae*) whose principle is not clearly known. This was the oldest assembly, but does not appear to have enacted legislation as such. It had a role in certain acts of private law (such as adrogation), and also possessed the important constitutional function, albeit reduced to a formality, by which the consuls were confirmed in the exercise of their power (*Lex curiata de imperio*). This may have played a part in the later juristic attempt to legitimise the law-making power assumed by the emperor (see section (c) below).

Plebeian statutes were enacted by an assembly composed only of the plebeians organised in tribes (*concilium plebis*). Voting was by groups as in the other assemblies, but the only magistrate competent to place a proposal before the plebeians was their own special representative, the tribune of the people. Because the patricians were not members of the assembly, there was doubt whether they were bound by its resolutions. This difficulty was removed by the Hortensian Act of 287 BC which provided that resolutions of the plebeian assembly should be binding on the whole people. After that date the plebeian assembly was the most widely used body for the enactment of legislation. Amongst its measures was the Aquilian Act of 287 BC which laid the foundation of the law on damage to property (see Chapter 5, section X(d) below).

(b) Resolutions of the senate

During the Republic the senate had no law-making power as such, although it had an influential role in the process by which proposals of the consuls, praetors, or curule aediles became law. Its consent was required before these magistrates could propose a law to the *comitia centuriata* or the *comitia tributa*, and it might itself instruct a magistrate to propose the enactment of a law. But actual law-making power was not acquired by the senate until the period of the Empire. The emperor Augustus (27 BC - AD 14) and his immediate successors sought to preserve the fiction that the Republic still existed, and so did not themselves assume formal legislative power. Yet they had to find some means by which their wishes could be translated into law. The early emperors found two ways to influence the legislative process. One was by the revival of the old assemblies of the people which had ceased to meet during the disruptive decades of the late Republic. The emperor in his capacity as consul or tribune could propose a law to whichever assembly he deemed appropriate. This practice itself is not known after the end of the 1st century AD. The other, more favoured, way was for the emperor to make a speech to the senate requesting it to approve a change in the law. The senate, with no choice in the matter, would then instruct the appropriate magistrate to insert in his edict (see section (d) below) the change requested by the emperor. For a long time there was an aversion to according the senate independent legislative power. This step was not achieved until the time of the emperor Hadrian (AD 117-38) in the first half of the 2nd century AD. Gaius, who wrote his *Institutes* around the middle of that century, lists the resolutions of the senate as a source of law, but adds that there was some doubt (G.1.4). What came in the end to be taken as the law was not the resolution of the senate as such, but the imperial speech outlining the proposal which was to be adopted. Once the emperors themselves were prepared to assume direct law-making power, the use of the senate as a law-making body ceased to be necessary. It did not function for this purpose after the end of the 3rd century AD.

(c) Imperial pronouncements
(J.1.2.6)

By the time of Hadrian the pretence that the emperors might not make law was dropped. This posed a problem for the jurists. Unless the emperors were to be seen as illegitimate tyrants, some justification for their legislative role had to be found. The late classical jurists found this justification in the fiction that the people had transferred to the emperor the legislative power which it had held since the expulsion of the kings in 509 BC. In a passage later used by Justinian in paragraph 6, Ulpian remarks (D.1.4.1 pr) that imperial decisions have the force of a legislative act because the people by a Regal Act has bestowed on the emperor all its own power and authority. This account raises two points of particular interest. Can this act by which the people transferred power to the emperor be identified, and why is it qualified by both Ulpian and Justinian as 'regal'?

The act referred to is most likely an act known to have conferred upon the emperor Vespasian (AD 69-79) a wide range of powers, although significantly these powers did not include the power to legislate. The act states that similar powers had been conferred upon Vespasian's predecessors. It is possible that it was modelled on the act of the ancient *comitia curiata* by which the executive authority of the higher magistrates had been ratified. The element of fiction in the juristic explanation of the power to legislate assumed by the emperors is discernible from the fact that it was this power precisely which was never transferred by the people.

The element of fiction assumes a larger role if we contemplate the possible significance of the word 'regal' by which the power-transferring act is named. This word may have been used merely because the act was one which concerned the powers of the emperor. But it is by no means impossible that the word had a different implication, namely that the emperor in virtue of the power-transferring act was now in the same position as the early kings were deemed to have been. The act of the *comitia curiata* concerning the power of the consuls may itself have been seen as the successor to a similar act in the regal period by which the people formally ratified the authority of the king. What Ulpian and Justinian may therefore have had in mind, reflected in the use of the word 'regal', is the placing of the emperor at the end of a long historical tradition, beginning in the regal period, and characterised by the transfer of power from the people to its rulers. In this way the actual exercise of legislative power by the emperors might receive the fullest possible authorisation.[8]

Imperial legislation, generally termed 'pronouncements' (*constitutiones*), took a variety of forms, the most important of which are enumerated in paragraph 6. A preliminary point to be borne in mind is that the emperors in exercising their legislative power generally relied on the advice of the officials closest to them, many of whom were skilled lawyers. The emperors had succeeded to the functions, and took the titles, of the principal republican magistracies. In that capacity they had the power to issue edicts which contained pronouncements upon the law that were intended to be of general applicability. An example is Caracalla's edict of AD 212 conferring citizenship upon the whole free population of the Empire. As supreme judge the emperor might hear cases either at first instance or on appeal. His decision was regarded not merely as binding on the parties to the case but, where appropriate, as establishing a precedent to be followed in similar cases. The emperor also supplied written answers ('writs') on queries as to the law submitted to him by officials or private persons. The imperial reply took one of two forms depending upon the source of the query. If an official sent a written request for instructions or for some special privilege, the answer took the form of a separate letter. Private individuals who sought favours or rulings on points of law could not send a letter to the emperor, but had to present him with a petition. His answer was then added at the bottom of the petition. Many important points of private law came to be decided in this way. One class of imperial pronouncements not mentioned by Justinian is the mandate. Mandates were administrative instructions issued by the emperor to his officials, often to provincial governors, and they came to lay down a comprehensive code for the administration of the provinces. Sometimes important rules of private law were introduced in this way, such as the rule that a soldier might make a valid will in any form without conforming to the normal requirements (see Chapter 4, section II(a) below).

Imperial writs and judgments gave rise to a serious problem of interpretation which Justinian himself notes in paragraph 6. In theory the position was clear. Sometimes a writ or judgment was intended to affect only the parties to the particular case, but sometimes it was intended to have a more general effect, to 'bind everyone', as Justinian says. The practical difficulty was that it was often unclear what intention was to be gathered from the particular writ or judgment. There seems indeed to have been an understandable tendency to treat any imperial decision as laying down a general rule. The post-classical period (4th and 5th centuries AD) in particular saw a considerable amount of legislation which attempted to prevent an unauthorised scope being given to writs and judgments originally meant to apply to one particular case.[9]

(d) Edicts of the magistrates
(J.1.2.7)

The edicts of the magistrates played a particularly important part in the development of the law once the strict and inflexible system of the actions in the law (*legis actiones*) gradually gave way to the formulary system during the 3rd and 2nd centuries BC (see Chapter 6, section II below). The essential point is that the appropriate magistrate under the formulary system had the power to determine whether an individual had a good cause of action or a good defence. Should he so determine, he would issue a formula in which the cause of action or defence was embodied, leaving it to the judge to determine whether the facts warranting the remedy or substantiating the defence could be proved. Each magistrate at the beginning of his year of office issued an edict in which he stated the grounds upon which he would grant a remedy or allow a defence. Although he had a scope for discretion in the exercise of his powers and might at any time introduce new remedies or defences, he tended on the whole to adopt the edict of his predecessor. Magistrates, not being themselves lawyers, relied upon the advice of the jurists (see section (e) below) who thus indirectly exercised an enormous influence on the development of the law. In this way a settled and consolidated body of law came to be established. The most important edict was that of the urban praetor, who had jurisdiction over disputes between Roman citizens and so had a fundamental role in developing substantive Roman private law through his control of remedies.[10] The foreign praetor, who had jurisdiction in cases involving foreigners, also issued an edict. Although little is known of its content, it was probably influential in developing the law of sale and other contracts used to facilitate commerce. Almost certainly considerable cross-fertilisation occurred between the urban and the foreign edicts. Finally the curule aediles, who had jurisdiction over markets and streets, issued an edict which regulated the sale of slaves and animals in the markets, and further imposed liability for harm caused by dogs or wild animals kept in public places.

The edicts operated as effective sources of law under the control of the magistrates mainly during the last two centuries of the Republic. At this time they represented the main vehicle for the development of the law. With the beginning of the Empire, changes might still be made to the edicts, but only with the approval of the emperor manifested in the form of a resolution of the senate. Around AD 130 the jurist Julian was instructed by the emperor Hadrian to 'consolidate' the edict of the urban praetor and cast it into a final, definitive form. From that time no further changes were permitted, and the edict ceased to be a mechanism for the introduction of new rules of law. The edicts of the foreign praetor and the aediles likewise appear to have been standardised by Julian, no change thereafter being possible.

(e) Juristic opinion
(J.1.2.8)

When considering the work and influence of the jurists,[11] the men learned in the law, to whose collective genius was due the creation of Roman private law during the first three centuries AD, we must at the outset make a critical distinction with respect to the kind of authority which their writings or their advice possessed. We have to distinguish between 'authority' in the sense of 'respect or influence' and 'authority' in the sense of 'recognised by the legal system as actually being a valid rule of law'. It is clearly the case that the jurists had an enormous, indeed critical, influence on the development of the law through the advice they gave the magistrates and the emperor, through the opinions they gave on legal issues referred to them by private individuals, and through the respect accorded their writings. But it is a different and more controversial question whether the Roman legal system contained a set of criteria specifying that any or all 'juristic answers' were to count as valid rules of law. The answer to this question concerns the exact nature of the 'right of replying' to which Justinian refers in paragraph 8. From Justinian's account, indeed, we might infer that jurists who were accorded the right of replying by the emperor were given the power to establish valid rules of law. But the correct understanding of this passage of the *Institutes* depends upon an appreciation of the history of the right of replying, a matter of some obscurity.

Before the time of Augustus (27 BC - AD 14) it is clear that the law formally accorded the opinions of the jurists no special weight. The respect they were given depended upon the standing of the particular jurist. Augustus introduced some changes in the status of juristic answers, and a further change was made by Hadrian. But the precise nature of the changes has remained something of a mystery, because the two principal texts which give an account of the changes are not sufficiently precise. The jurist Pomponius (a contemporary of Gaius), in the course of a general survey of legal history (D.1.2.2.49), states that before the time of Augustus the right of giving opinions in public was not granted to individual jurists. However, Augustus, so that the law might possess greater authority, provided that opinions might be given under seal with his own authority. Subsequently the privilege began to be sought. Hadrian is cited as also making a ruling on the right of replying, but it is not clear from the somewhat garbled account in the extant text of Pomponius what this exactly was.

The second piece of evidence is supplied by Gaius in his *Institutes* (G.1.7). The first part of this text, adopted by Justinian, defines juristic answers as the opinions of those entrusted with the task of founding the law. Then comes a significant statement to the effect that, where such opinions agree, the rule that they express has the force of an act. On the other hand, should they disagree, the judge may select which one he prefers. That concordant juristic opinion shall have the force of an act is attributed to a writ issued by Hadrian.

The chief problems arising from these two texts are: (1) did the right of replying granted by Augustus confer actual binding force upon the opinions of a jurist to whom the privilege was granted; (2) did Hadrian's writ confer binding force only on opinions obtained in a particular case, or did it go further and endow such opinions with binding force as precedents for future cases, or even make binding the concordant opinions to be found in the writings of the jurists; and (3) what was the relationship of Hadrian's writ to the privilege granted by Augustus? No agreement has been reached by scholars on these questions. Some accept the view that Augustus in conferring the right to make public replies on selected jurists did no more than permit those jurists to give their opinions with the extra weight of his own prestige. A judge or magistrate who received an opinion from a jurist so favoured would undoubtedly follow it, since it came with the approval of the emperor. Nevertheless, technically the opinion was not binding on the judge; it did not in itself embody a valid rule of law.[12] This view has a certain degree of plausibility, since no text expressly says that Augustus conferred the force of law on the opinions of selected jurists, and since he himself made a particular point of not assuming the power to legislate.

Against this view is a strong current of authority to the effect that Augustus did in some sense make binding the opinions of the privileged jurists.[13] Scholarly assessment of Hadrian's writ has been equally diverse. Some writers assume that Hadrian for the first time made juristic opinions binding, provided that they concurred, but only in respect of the particular case for which they were given. Others hold that Hadrian also provided that the opinions of jurists with the right of replying should constitute binding precedents where they concurred, and even that congruent opinions expressed in the juristic literature should have the force of law.

The view accepted here is that Augustus conferred more upon the opinions of selected jurists than merely the prestige of his approval. It can reasonably be inferred from Gaius' version of Hadrian's writ that, prior to Hadrian, the opinions of the jurists in some sense were regarded as binding. From Gaius' words it looks as though Hadrian was concerned to regulate opinions which had already become binding rather than to introduce for the first time the principle of a binding opinion. Nor is it easy to see the point of Augustus' intervention unless he did make binding the opinions of certain eminent jurists.

If we assume that Augustus did make the opinions of those jurists accorded the right of replying binding upon the judge or magistrate for the particular case in which they were given, it is reasonable to suppose that between the time of Augustus and that of Hadrian, more than a century later, further changes took place, either with the authorisation of the emperor or just as a matter of practice. What is very likely to have happened is that the opinions of jurists with the right of replying given for one case came to be cited as binding precedents for later, similar cases. It may even have been argued that any formal expression of opinion by such a jurist possessed binding force. By the time of Hadrian some regulation of what had become a confused position was necessary. A judge or magistrate might be confronted by a series of conflicting opinions each asserted to be binding, since given by a jurist with the right of replying. How was he to choose between them? On the approach taken here Hadrian regulated the position exactly in the way described by Gaius. He confirmed that opinions were binding if they concurred, but gave the judge or magistrate the power to choose between them where they were not.

There is almost no information in the sources as to which jurists received the right of replying. Pomponius states (D.1.2.2.50) that Sabinus, one of the most influential jurists of the 1st century AD, received the right from the emperor Tiberius (AD 14-37). No other jurist is ever expressly identified as having received it, but in view of the general observations in the texts of the problems arising from juristic opinions, we may conjecture that by the time of Gaius, at any rate, most eminent jurists enjoyed such a right. We may also conjecture that, whether or not this had been the original intention of Augustus, any formal opinion of a jurist with the right of replying was treated as a binding statement of the law.

1. Sources

Whatever the precise nature and history of the right of replying, later ages came to regard the writings of the classical jurists as a *corpus* which itself constituted the law. Any opinion was authoritative in the sense of possessing binding force. However, the degree of binding force had to be relative, since in so vast a body of legal literature discrepant opinions might readily be found. This raised the issue, to the opinions of which jurists should the greater authority be accorded in the event of a conflict? The problem was solved by the Law of Citations promulgated by the emperor Theodosius II (AD 408-50) in AD 426. Five of the classical jurists were identified as those whose works possessed the greatest authority. Should there be a conflict between them, the majority view was to prevail. In the event of the opinions on each side being evenly balanced, the side which enjoyed the support of the jurist Papinian was to prevail. Although much criticised for providing a mechanical text, the Law of Citations established a regime arguably not so very different from the Anglo-American system of binding precedent, under which the decisions of certain courts are treated as more authoritative than those of others.

IV. Custom
(J.1.2.9, 11)

In title 2, paragraph 9 Justinian takes up a distinction made by Ulpian in the *Digest* (D.1.1.6.1) between written and unwritten law, but elaborates it specifically by adducing custom as an example of unwritten law. Custom can be taken to be a source of law in two principal senses. First, one may speak of custom as having been a source of law from an historical perspective: in the past a number of rules forming part of the legal system had been observed in customs or usage of the people, and these rules were either gradually accepted as law or made into law through some particular act such as legislation or adoption by the courts. For example, the *Twelve Tables* are regarded as largely being a codification of existing custom and practices. Furthermore, the magistrates in their edicts will have drawn upon contemporary practices, especially perhaps in the field of commercial relations, in framing new rules of law. The jurists, too, may have been influenced by traditional practices in reaching an opinion on a point of law. From this perspective, when one states that custom is a source of law, one is simply making a factual assertion that customary practices have been absorbed into the legal system through some particular mechanism, be it legislation, magisterial edict, or juristic answer. As a consequence of this process of absorption what is previously unwritten practice has become written law. But one can hardly regard a practice, prior to its absorption into the legal system, as having been unwritten law, unless one assigns to 'law' in this context the loose sense of 'that which is generally and habitually followed'.

The second sense in which one speaks of custom as a source of law is rather different. Here one is not making an historical statement, but pointing to a set of criteria, recognised by the legal system, by which custom, habitual practices, are identified as being rules of law. The problem for Roman law is to determine whether custom is ever recognised as a source of law in this sense, and, if so, to define the criteria by which it was identified. The classical jurists on the whole are silent on the matter. Julian does state (D.1.3.32) that immemorial custom has the force of a statute, on the ground that such a custom, like an act itself, is binding on account of the approval of the people. If this statement is taken to be authentic and not, as is often assumed, to be an interpolation, it may be construed in a number of ways. Perhaps Julian was merely making an historical assertion to the effect that in the past, immemorial custom has been absorbed into the legal system. This construction is not plausible, since Julian's language definitely seems to advert to custom as a source of law in the second sense, that is, identifying immemorial custom as itself being law with the same force as a statute. If this is his meaning, he unfortunately gives no clue as to the type of custom he had in mind or the criteria by which it was to be identified as law. In particular, given the size and diversity of the Empire in his day, what general customs accepted by the people at large could he have had in mind? Was he thinking only of those accepted by Roman citizens? Could he, for example, have had in mind an institution such as the power exercised from immemorial times by the head of the family over his children? Or could he have been thinking of a much more important and practical issue, the status of local custom which might be followed in particular areas of the Empire in derogation of the normal (written) law? In view of the breadth of his language, Julian can perhaps be taken as making a general statement of jurisprudence without thinking particularly of practical consequences. He states the proposition that logically the consent of the people makes a custom, just as a statute, into law, but is not concerned with the more practical question: which precise customs are deemed, through the consent of the people, to be law?[14]

We may take the same approach to the statement in paragraph 9 that long-standing custom is like legislation. It may be understood as a statement of jurisprudence which elucidates the concept of unwritten law at the general and theoretical level; or it may taken as a statement of practical import for the operation of the Roman legal system. On the latter construction, Justinian is probably drawing attention to the significance of local and regional customs. They are to be given effect as though they are statutes. The fact that he speaks carefully of custom as being like, rather than as having the same force as, legislation, perhaps carries the implication that he would not have regarded such a custom as prevailing over a contrary act.

In paragraph 11 Justinian makes what appears to be another reference to the legal effect of custom. Contrasting the law which each state establishes for itself with the law established by divine providence, he observes that the former may be changed either by the tacit consent of the people or by a later act. Is this a reference to the fact that Roman law operated a doctrine of desuetude according to which a statute might cease to be valid if it could be proved that it had not been observed for a certain period of time? If so, it seems to conflict with a ruling of the emperor Constantine (AD 312-37), who had declared in a promulgation of AD 319 that, although custom possessed considerable authority, it might prevail against neither reason nor legislation (C.8.52.2). However, these apparently conflicting statements may perhaps be reconciled if we again bear in mind the distinction between theory and practice. Constantine may have been thinking of the possible efficacy of local or regional custom; while being prepared to accord it a certain authority, he was not willing to permit such a custom to prevail over a contrary statute. On the other hand, Justinian, like Julian who states explicitly that law may be abrogated by desuetude in virtue of the tacit consent of the people (D.1.3.32.1), may have been reflecting on the logic of the consent of the people in relation to law. If the consent of the people ultimately accounted for the validity of legislation, it logically seemed to follow that withdrawal of that consent would entail invalidity. Nevertheless, such a doctrine, although acceptable in theory, could have awkward consequences in practice. As we know from the discussion in the *Institutes* of imperial pronouncements, the people was deemed to have transferred its legislative power to the emperor. From this it could also be deduced that, once legislation was passed or endorsed by the emperor, the people no longer had the power to deprive it of its validity. Generally, it should be noted that neither the jurists nor the emperors, on the existing evidence, make any attempt to establish the exact criteria by which a practice is to be identified as a custom possessing legal force, or a statute is to be seen as having fallen into desuetude. This reinforces the point that general statements about the legal effect of custom or the tacit consent of the people were not understood as having particular relevance to the actual operation of the legal system.

V. Divisions of Private Law
(J.1.2.12)

In title 2, paragraph 12 Justinian delimits the main divisions of the private law with which he proposes to deal. The law of persons (Chapter 2 below), sometimes described as the law of status, comprises much that in modern law is termed family law. The law of things examines the rights and obligations generated by the use and exploitation of economic assets, covering what in modern law is termed the law of property (Chapters 3 and 4 below), the law of contract and unjustified enrichment (Chapter 5, I-VI below), and the law of delict (Chapter 5, IX-X below). The law of actions

(Chapter 6 below) covers the nature of the remedies by which rights are enforced, what in modern law is often termed the law of procedure.

Select Bibliography

Jolowicz, H.F., and Nicholas, B., *Historical Introduction to the Study of Roman Law* 3rd ed. (Cambridge, 1972) chs 2, 5, 6, 21, 22, 27.
Kunkel, W., *An Introduction to Roman Legal and Constitutional History* 2nd ed. trans. J.M. Kelly (Oxford, 1973) chs 1, 2, 5-8.
Robinson, O.F., *The Sources of Roman Law: Problems and Methods for Ancient Historians* (London, 1997) chs 1 and 2.
Tellegen-Couperus, O., *A Short History of Roman Law* (London, 1993).

Notes

1. On the concept of public law see in particular D. Johnston, 'The General Influence of Roman Institutions of State and Public Law', in D.L. Carey Miller and R. Zimmermann (eds), *The Civilian Tradition and Scots Law: Aberdeen Quincentenary Essays* (Berlin, 1997) pp. 87ff; P. Stein, 'Ulpian and the Distinction between ius publicum and ius privatum', in R. Feenstra et al. (eds), *Collatio Iuris Romani: Études Dediées à H. Ankum* (Amsterdam, 1995), pp. 499-504. On the distinction between public and private law see further J.A.C. Thomas, *The Institutes of Justinian* (Cape Town, 1975) pp. 3-4.

2. For the distinction between 'historical' or 'material' sources and 'formal' or 'legal' sources see H.L.A. Hart, *The Concept of Law* (Oxford, 1961) p. 246.

3. Cf. the remarks in Nicholas, *Introduction* pp. 55ff.

4. Cf. M. Kaser, *Ius gentium* (Cologne, 1993) p. 60n234.

5. Kaser, ibid. pp. 4ff.

6. D.1.1.7 (Papinian, *Definitions*, book 2), summarised below. See also Chapter 6, section V.

7. It should be remarked that the most famous Roman acts, the code of the *Twelve Tables*, were produced not by an assembly of the people, but by a body of special commissioners elected by the people.

8. Cf. M. Kaser, *Ausgewählte Schriften* I (Naples, 1976) p. 182n4.

9. For discussion see J.B. Moyle, *Imperatoris Justiniani Institutionum* 5th ed. (Oxford, 1912) pp. 104-5.

10. This edict has been reconstructed by O. Lenel, *Das Edictum Perpetuum* 3rd ed. (Leipzig, 1927).

11. Generally on the lives and writings of the classical jurists see F. Schulz, *History of Roman Legal Science* (Oxford, 1963) and W. Kunkel, *Herkunft und soziale Stellung der römischen Juristen* 2nd ed. (Cologne, 1967).

12. See, for example, F. de Zulueta, *The Institutes of Gaius* II (Oxford, 1953) pp. 20f; H.F. Jolowicz and B. Nicholas, *Historical Introduction to the Study of Roman Law* 3rd ed. (Cambridge, 1972) pp. 359ff; Thomas, op. cit. above note 1, p. 12.

13. See, for example, H. Honsell, T. Mayer-Maly, and W. Selb, *Römisches Recht* 4th ed. (Berlin, 1987), p. 27; Kaser, *RPR* I, pp. 210-11.

14. Cf. Kaser, *RPR* I p. 196; II p. 57.

2. Persons

O.F. Robinson

Justinian's *Institutes* follow those of Gaius (G.1.8) in dealing with persons after the introductory titles or sections on the nature and sources of law. For both, although less markedly for Justinian, the law of persons was not so much about dynamic legal relationships – as modern lawyers are inclined to see it – but about status.[1]

I. Slavery
(J.1.3 - 8)

The fundamental distinction between human beings was that they were either free persons or slaves (J.1.3 pr; G.1.9); no other distinction had the same legal weight. The institution of slavery was universal in the ancient Mediterranean world. Gaius takes it completely for granted and slavery still existed in the Christian empire of Justinian. Admittedly, Justinian refers to liberty as man's 'natural' ability to do what he wants (J.1.3.1), and he held that slavery, although an institution of the law of all peoples (*ius gentium*), was at the same time contrary to the law of nature (J.1.3.2) (see Chapter 1, section II(b) above). Moreover, free birth was something which could not be taken away (J.1.4.1); subsequent misadventure could not affect this right beyond price. Nevertheless, slavery continued into the Byzantine Empire, while in the Germanic settlements of the West it slowly died out.

(a) Enslavement
(J.1.3 - 4)

Justinian explains the institution of slavery as deriving from warfare. Captivity in the ancient world automatically implied enslavement. Romans who were captured became slaves, just as did those whom the Romans themselves captured. (What was unusual at Rome among ancient Mediterranean states was that slaves who were freed normally became citizens.) In Roman law there were other causes of enslavement: in early law, for example, it was a consequence of evasion of the census (with intent to avoid military service) or of judgment debt; in later law, the sale of oneself in order to share in the profits of the sale or condemnation to death in a criminal court were among these causes. For gross ingratitude, a freedman's manumission (release from slavery) could be revoked by his patron. There was also a Claudian Resolution of the Senate (of AD 54) whereby a female Roman citizen could cohabit with a slave with his owner's consent but any offspring would be born slaves (G.1.84). On the other hand, if the cohabitation was against the owner's will but the woman persisted in it despite his warning, she herself could be enslaved (G.1.91), although we learn elsewhere that if she were in paternal power her father could prevent this (Paul's *Sentences* 2.21A.9). This resolution of the senate probably[2] originated in the attraction brought to bear by imperial slaves, *servi Caesaris*, who during the early Empire exercised a degree of political power, and in the emperors' desire that these men's sons should be bound to follow them in the imperial service. It was repealed by Justinian himself as unworthy of the times (J.3.12.1).

The major source of slaves, however, apart from captivity (which ceased to be at all common after the early Empire), was birth to a female slave; outside Roman law marriage, the rule (which stemmed from the law of all peoples) was that a child took the status of its mother. But the texts stress that if the mother, although a slave at the time of the birth, had been free at conception or at any time during her pregnancy then the child would be born free (J.1.4 pr). The lawyer's habit of citing authority is neatly illustrated in this text, where Justinian appeals to the authority of the jurist Marcellus, who lived in the latter part of the 2nd century but was not as prominent as, say, Julian earlier or Papinian later.

(b) The legal condition of slaves
(J.1.3.5; 1.8.1 - 2)

Justinian remarks that 'the legal condition of all slaves is the same' (J.1.3.4). There were a few technical exceptions to this generalisation; we shall look at them at the end of this section. In practice, however, there were enormous differences.

A slave was a thing, subject to commerce like any other thing. He could own nothing; he had no rights; he could not make any legally binding commitment, although he could perform some legal acts; he could not marry and, legally speaking, he could have no relatives. A slave could be bought and sold, lent and hired out, deposited and pledged, entirely at his owner's will. He was an item of property, probably valuable property, but property in the same way as a horse. The rigour of this principle was inevitably modified by his human ability to think and speak, which made him a useful instrument of his owner. The praetor came to grant actions against the owner in various circumstances where the slave had an explicit or implicit authorisation to bind that owner (G.4.69 - 74a). Further, a slave had always the potential to become a free man and a citizen, and so he could incur a natural obligation. He was also personally responsible (although not legally liable, because as a slave he could not be sued) for his delicts; if the delict had not yet been purged, any new owner of the wrong-doing slave acquired this legal liability along with ownership, and if the slave were freed he himself could then be sued. The maxim to explain this sort of liability was 'noxa caput sequitur' (see J.4.8; G.4.75 - 79). The slave was also held responsible for the crimes he committed.

Moreover, although the legal status of all slaves was technically the same, an intelligent and able slave who was trusted by his owner with the management of a business or an estate could become *de facto* a rich man, himself 'owning' many slaves (known as *servi vicarii*) and enjoying roughly the same standard of living as citizens in the upper classes of society. Less fortunate slaves could still, by their owners' grace, save their earnings and their tips – included in their personal fund, *peculium* – with which they might perhaps 'buy' their freedom, or which they could hope to take with them into freedom. Other slaves worked in gangs, beaten into submission, treated worse than animals. It was to remedy such abuse of rights by owners that anti-cruelty legislation was passed by the emperor Antoninus Pius (AD 138-61), a Stoic by philosophic belief (J.1.8.1 - 2; G.1.52 - 53). In the Republic the censors had sometimes condemned men for unbecoming behaviour to their slaves; the emperor Claudius (AD 41-54) ordered a sick slave who had been abandoned by his owner to be free (though not a citizen), and the emperor Hadrian (AD 117-38) restricted an owner's right to kill his slaves. Nothing of this, however, gave the slave himself any 'right'; all he could do was implore mercy.

There were also a few slaves whose status was specially distinguished from that of the normal slave in private ownership. Penal slaves (*servi poenae*) were those, free or slave, who had been condemned to death or some equivalent punishment, such as labour in the mines. They did not even have a derivative capacity, i.e. there was no one whom they could represent, since they were without an owner, being enslaved to their punishment; imperial pardon was their only hope. Slaves in the service of the state (*servi publici*) in republican Rome, and in the municipalities then and thereafter, were in public ownership and operated as minor functionaries; they could represent their city up to a point. *Servi Caesaris*, imperial slaves, might in the early Empire exercise considerable power as more than minor civil servants, but their special position was overtaken first by imperial freedmen – themselves at a later stage, so to speak – and then by men of equestrian status. These three groups, however, were never more than a tiny proportion of the slave population.

(c) Manumission
(J.1.5 - 7)

Justinian mentions different methods of manumission, but for him any specific act with this intention was effective, unless it was forbidden for some reason. All valid manumissions made the former slave a citizen (J.1.5.3). The only restrictions mentioned in his *Institutes* are on manumission which defrauded creditors and on manumission by will attempted by those under 17. (The freeing of a slave so that he could be compulsory heir to an estate which was bankrupt was not viewed as fraudulent, because it was better for the creditors that there should be someone to act as executor: J.1.6.1.) Both these restrictions stemmed from modifications of the Aelian-Sentian Act of the early Empire; it had denied the power to manumit to owners under 20, even though testators could dispose of all other kinds of property by will from the age of puberty (fixed at 14 for boys and 12 for girls: J.1.22 pr; G.1.196. See section V(d) below). Justinian repealed the rest of the legislation on manumission and inferior classes of freedmen which had been passed in the very early Empire, and with which Gaius was much concerned. But Justinian's claim to have restored the state of the law in the Republic was not quite accurate; in the Republic only three formal modes of manumission had granted legal freedom, and an informal grant had produced an insecure status in which the former slave was withheld by the praetor from the former owner if servile duties should be demanded of him, but was not otherwise protected by law.

In the time of Gaius there were still only the three methods of manumission which made a citizen (G.1.17): by rod, by will, and by enrolment in the census (which seems to have been last held in Italy under the emperors Vespasian (AD 69-79) and Titus (AD 79-81) in the 70s). Informal manumission had, however, been regularised by a Junian Act, passed under either the emperor Augustus (27 BC - AD 14) or Tiberius (AD 14-37),[3] which created the status of Junian Latin.[4] A Junian Latin, normally referred to by Gaius simply as a 'Latin', was a foreigner, but a privileged one (G.1.22 - 24). His person was inviolably free but, if he failed to attain citizenship, his estate at his death reverted to his former owner and did not descend to his children; as Justinian says (J.3.7.4): 'they were free during their lives but gave up their freedom with their lives.' There were many ways by which a Junian Latin could become a citizen (G.1.28 - 35; cf. G.1.66 - 73, 79 - 81, 167), and it seems likely that only relatively few male Latins (it was certainly harder for females) could not attain citizenship, if they wished.

Other legislation on manumission was undoubtedly passed under Augustus, that is, the Fufian-Caninian Act of 2 BC and the Aelian-Sentian Act of AD 4.[5] The first dealt with manumission by will (G.1.42 - 46), restricting the proportionate number of slaves an owner could manumit. At a time when the Empire seemed still to be expanding, and when border war was fruitful in booty, including slaves, it may have seemed sensible to put some control on the numbers of former slaves admitted to the citizen body, and in particular on those freed by will. By this kind of manumission a man could ensure a great show for his own funeral, to the loss of his heirs, who inherited a diminished estate, and at the possible expense of the state, which needed to find employment for the urban poor.[6]

The Aelian-Sentian Act imposed limitations of age on both owner and slave; if the owner was under 20 or the slave under 30 there was normally no fully valid manumission. Patience could cure the age restrictions, and there was also a procedure by which exceptions could be made for good cause (J.1.6.4; G.1.18 - 20, 38 - 39). As well as prohibiting manumissions which defrauded creditors (and patrons in cases where the owner was himself a freedman), the act created a special status of capitulated alien (*dediticius*) where a slave was freed who had been the subject of a serious criminal conviction or had received severe domestic punishment. The capitulated aliens of the Aelian-Sentian Act were truly free, but they were less privileged than ordinary foreigners, for example, in where they could live (G.1.13 - 15, 25 - 27). Their status, which Justinian describes as having fallen out of use 'long ago' (J.1.5.3), probably disappeared with the absorption of slaves into the ordinary procedures of the criminal law and their (theoretical) removal in matters of serious crime from the domestic jurisdiction of their owners. Well before the end of the Republic foreigners were regarded in law as having rights of their own and as sharing with the Romans in the law of all peoples. However, 'foreigners' were all those who were not Roman citizens, including subjects of the Empire as much as those from outside its borders. Only in the early third century, by the Antoninian enactment of AD 212 (Caracalla's edict), did virtually all free inhabitants of the Empire become citizens. This is why Gaius pointed out that the clause of the Aelian-Sentian Act on defrauding creditors also applied to manumission by foreigners, although the rest of the measure did not (G.1.47).

The main point was clear. In both classical and Justinianic law a freed person was free. Some social stigma might be attached but, once freed, a man or woman was on the other side of the greatest divide. There were certain disabilities for freedmen in public law, and usually they owed certain duties to their patron (their former owner), and perhaps his family or heirs, but a citizen freedman was a citizen with normal rights, except where those rights were specifically restricted. For the Junian Latin his personal freedom was safe, and even if he died before reaching citizenship his children by any woman not a slave might be able to benefit from his potential upward mobility (G.1.32).

II. Family Authority

(J.1.8 - 9, 11; G.1.108 - 123, 135 - 141)

The second stage in the classification of persons was that some were independent, while others were subject to someone else's power. Both Gaius and Justinian here mention slaves, as we have seen, as being by the law of all peoples subject to the authority of their owners. But among free persons the principal aspect of dependency was subjection to paternal power (*patria potestas*). In classical law and earlier there were two other kinds of dependency: marital subordination (*manus*: G.1.108 - 115b, 136 -

137a) and bondage (*mancipium*: G.1.116 - 123). We shall deal with these after paternal power.

(a) Paternal power acquired by birth (J.1.9 pr, 2 - 3)

This gave the father authority over his legitimate children for his lifetime. Marriage, which we shall consider shortly, was mainly viewed in Roman law as the normal avenue for the creation of paternal power. This is why Justinian's title on marriage comes between those on family authority and on adoption, a second way by which paternal power could be achieved. This is also why Gaius devotes so many sections (G.1.65 - 96) to the complications caused by parties to a marriage having different status, or being mistaken as to each other's status.[7] He deals there with the acquisition of paternal power, in cases which involved some or all of the parties concerned – including the children – entering in due course into relationships of the state law with each other, whether they had begun as citizens or Junian Latins or foreigners. All these special cases were relevant in classical law but no longer in Justinian's, when free men were either citizens or foreigners, and foreigners were indeed from abroad, not fellow-inhabitants of the Empire.

Gaius' description is straightforward, if somewhat lengthy. It is made clear that the rules applied to daughters as well as sons (G.1.72), and that the concept of lawful marriage and legitimacy could be found among foreigners (G.1.77). The normal rule of the law of all peoples, that a child followed its mother's status, was affected by various statutory provisions: the Minician Act (G.1.78 - 79); resolutions of the senate where the peculiar status of a Junian Latin might logically have resulted in a state-law marriage in which the children would follow the status of their father (G.1.80); or the otherwise unknown *lex* (which may have applied to a particular province) on the offspring of free men and female slaves (G.1.85 - 86). Gaius also stresses that legitimate children acquired their status at the time of their conception, but that others took it from the time of their birth (G.1.89), and he links this with cases when a wife suffered banishment from home and hearth (G.1.90).

The Roman institution was different from the paternal power found among many races (G.1.55) in that it did not end when the child, male or female, came of age; a traditional Roman problem of etiquette was the question of precedence between a man who was consul (the highest office in the Republic) and his father. Because family authority was lifelong, the biological father did not necessarily exercise power over his own children since he might still be in the power of his own father, who was thus head of the family for his children and his sons' children, and indeed for the children of his sons' sons.

Paternal power originally conferred a literal power of life and death over the children subject to it; this had been much modified by Justinian's time,[8] when a father was restricted to reasonable chastisement, but it survived in his right to recognise as his own his newborn children, or to refuse recognition and either expose or sell them (sale was preferred to exposure, which the Christian emperors tried to stamp out).[9] Further, within the family unit, the father was, in principle, the only one who had proprietary capacity; his children could not own anything, any more than slaves could, although the personal fund of a son on military service was, for good pragmatic reasons, treated as though he did actually own it.[10] This principle too was much modified in Justinian's law. Again, a father's consent was necessary for his children's marriage. Since the exercise of power went through the male line, there followed the slightly odd result that it was more important for a son than a daughter to have consent to marry. Justinian makes the point (J.1.10 pr) in relation to a father incapable of giving consent by reason of insanity. Since a father had no power over his daughter's children – this was exercised by the husband or by his father – her insane father could be ignored or his consent presumed, since he would not be responsible in any legal matter for any offspring; one supposes that in the case of sons it had been necessary to assume a fiction of a lucid interval. Finally, the father exercising authority was liable for the delicts of those in his power.

(b) Paternal power acquired by adoption
(J.1.11; 1.12.8)

While paternal power was normally acquired through marriage, it could also be created by the institution of adoption. The whole point of adoption was succession through the continuation of the agnatic line, that is, descent traced through the male. To some extent the history of the law of persons is the history of the decline in the significance of the agnatic family; it was still fairly important in classical law, which is why Gaius wrote flatly (G.1.104) that 'women cannot adopt by any method, because they do not have power even over their real children'. This was modified somewhat by the emperor Diocletian (AD 284-305), following whom Justinian says (J.1.11.10), after quoting Gaius: 'But by imperial favour they are allowed to adopt to make up for the loss of their own children.' The only effect was to give such children a right of intestate succession to the woman. The institution of adoption was considerably reduced in importance by Justinian himself. For Gaius it had meant that the adopted person ceased to be in his old agnatic family and became fully a member of his new one, although the cognatic ties of blood relationship with his old family were, of course, unaffected. For Justinian it did not involve this complete transformation.[11]

> But now, by our own pronouncement, when a real father allows a son in his

> authority to be adopted by an outsider the rights implicit in the real father's authority are not affected and do not pass to the adoptive father. The son does not enter into the adoptive father's authority, although we have conceded him the right to succeed on intestacy. (J.1.11.2)

Only when the adoptive father was the adopted person's ascendant did the old legal effects take place; a child could in this way move into the same power as his mother.

There were two forms of adoption, the 'adrogation' of someone not in paternal power, and the 'adoption', in the narrow sense, of someone still subject to another's authority. Adrogation was a much more serious affair, particularly in classical law, for it meant the end of the adrogated man's own agnatic family. Even in Justinian's law it had more fundamental consequences, for the adrogated man continued to take any children or grandchildren he had in his power, as well as any property he owned, into the authority of his adoptive father.

At both periods adrogation required a legislative act. In pagan Rome it was carried out before one of the assemblies of the people under the presidency of the chief priest (*pontifex maximus*); in the Christian Empire, and even before, it was performed by imperial enactment. In both cases there was an inquiry into its necessity, particularly careful when it was proposed to adrogate a boy below puberty; in such a case, under Justinian's law, the adopting father had to give security to a court official so that in the event of the boy's death his original family would benefit – no scope for wicked adoptive fathers to commit murder. Moreover, he was not allowed to cast him off by emancipation (see section IV(c) below) while retaining the property the boy had brought with him (J.1.11.3). Since the purpose of adrogation was to continue the agnatic line where the adopting father had no children, it was pointless to adrogate a woman since she was by definition the end of her own family line. There was, however, no other way in classical law for an illegitimate daughter to enter her father's family, and this may partly explain the introduction of adrogation by imperial enactment. This form, moreover, became much more convenient when, after the expansion of the citizenship, most 'Romans' lived elsewhere than in Italy.

Adoption in the narrow sense in classical law used the ceremony of mancipation (see section IV(c) below) to release the child, male or female, from paternal power, following which the adopting father made a collusive claim before the praetor that the child was within his authority, whereupon the magistrate adjudged the child to him. Because of the lifelong nature of paternal power, the adopted person need not be a child. There was, however, a rule that the natural relationship should be imitated and so it was forbidden that a younger adopt an older man; Justinian lays down specific regulations. Nature did not need to be followed completely in that a natural eunuch was allowed to adopt; however, someone who had been castrated was not able to do so.[12] Justinian simplified the procedural requirements for adoption in the narrow sense, merely demanding a declaration before the relevant court. Even in Justinian's time, adoption was more than simply giving the adopted person rights of succession on intestacy to his adoptive father. This is made clear by the fact that adoption need not be of someone as son or daughter. A man might wish someone to come within his paternal authority as a grandchild, or even great-grandchild; however, the rights of a potential holder of this authority, e.g. a son, were safeguarded (J.1.11.7). Nevertheless, a grandfather could give his grandchild away in adoption without the consent of his son, the child's biological father.

Adoption by will was never known to Roman law.[13] If a man wrote such a thing in his will, it merely imposed a moral duty on the heir to take the testator's name. In the most famous example, Augustus' adoption by Julius Caesar (d. 44 BC), the adoption was ratified by the assembly retrospectively. If a slave was adopted in a will, this was held to give him freedom and make him heir, but it did not bring him into the testator's agnatic family; if a man wished truly to adopt a slave he must first free him and then adrogate him.

(c) Paternal power acquired by legitimation
(J.1.10.13)

Illegitimate children were by definition born independent of paternal authority (J.1.10.12). They were related by blood, cognatically, to their mother, who had a duty to aliment them, but they had no legal relationship to their father. We have already referred to Gaius' discussion (G.1.65 - 96) of what was in effect legitimation in classical law, when statute intervened to correct mistakes as to status (section II(a) above). In later Roman law, under Christian influence, it became possible for children to be legitimated (J.1.10.13) simply by the subsequent marriage of their parents, where there had been no impediment to such marriage at the time of their birth, such as the mother being a slave or one parent being married to someone else. Another means of legitimation arose from the difficulty of getting people to serve on municipal councils in the later Empire. A son could be legitimated when his father presented him to become a member of such a council; a daughter could be legitimated if, suitably dowered, she was married to a municipal councillor.

(d) Marital subordination
(G.1.108 - 115b, 118, 123, 136 - 137a)

We shall consider the ways by which a woman came into marital subordination (*manus*) under the heading of marriage (section III below). Here we shall briefly detail its effects as one form of family authority. *Manus* came about through a form of marriage, known at least as early as the *Twelve Tables* of the mid-5th century BC, in which the woman went into the power of her husband. All the property she had at the time of the marriage, and any acquisitions she might afterwards receive, became the property of her husband. In this way and in her rights of succession she stood in the position of a daughter to her husband. This was the true version of *manus*. However, it was also used as a contrivance, under a kind of trust, by which a woman, during the centuries when independent adult women were under the restriction of guardianship (section V(f) below), might change her guardian or make a will (G.1.114 - 115a). Justinian makes no mention of the institution, which seems to have been obsolescent even before the beginning of the Principate, although it may have been revived for sacral purposes.

(e) Bondage
(G.1.116 - 123, 135 - 135a, 138 - 141)

Bondage, the state of being *in mancipio*, was even in classical law almost always a temporary state, a device used in bringing about some other change in status, such as adoption (or emancipation; see section IV(c) below). The ceremony which created it, that of mancipation by bronze and scales, is described fully by Gaius (G.1.119). It had general applicability to those objects in the law of property which needed formal conveyance (*res mancipi*; see Chapter 3, section I(b) below) as well as to familial purposes. The one case where bondage was more than temporary was consequent upon the noxal surrender of a son for some delict. The free status of the one in bondage was preserved; he could contract a valid marriage and have legitimate children. Any acquisitions, however, like those of a slave, were the property of the one who held him in his power. Selling or delivering someone into such bondage was a function of paternal power, but release was by manumission as used for a slave, although not subject to the statutory limitations. The ceremony of bronze and scales had died out long before Justinian, who makes no mention of it.

III. Marriage
(J.1.9.1; 1.10; G.1.108 - 113, 115b, 136 - 137a, 148)

(a) Free marriage
(J.1.9.1; 1.10)

Marriage does not seem to need very much comment. The normal form of marriage both for Gaius and for Justinian was the so-called 'free' marriage, where the wife did not fall under the authority of her husband and her property remained separate from his. It was not very different from our present European version, but far removed from the institution developed in feudal and Germanic custom which remained normal until well into the 20th century.[14] Because the principal legal importance of marriage was the creation of paternal power and the perpetuation of the agnatic family, very little is said in either *Institutes* about its normal concomitants such as betrothal and dowry, or the actual relationship between husband and wife, or indeed divorce.[15] Some years after the publication of the *Institutes*, Justinian consolidated and reformed the law on marriage (Nov.22 (AD 536); 117 (AD 542); see also Nov.97 (AD 539); 98 (AD 539); 140 (AD 566)).

The first two requirements of marriage were age and consent (J.1.10 pr). The lawful age of marriage was puberty, which came to be fixed at 12 for girls[16] and 14 for boys. An apparent marriage entered into before puberty ripened into marriage on attainment of the necessary age if there were no other impediment. Consent was needed from both the parties to a marriage, wife as well as husband, although if she was in her father's power he concluded the marriage on her behalf. (This was so at least in theory: the marriage of Cicero's daughter to Dolabella was brought about by her and her mother, against Cicero's own wishes.) But because the wife's father was not responsible for the grandchildren, by not prohibiting the marriage he was apparently presumed to consent. The cessation by either party of the continuing consent to the marriage (*affectio maritalis*) was, in principle, sufficient for divorce in classical law.[17] In the Christian Empire causeless divorce was penalised, but effective.

The third requirement was the capacity to marry according to state law (*conubium*). This capacity was dependent on the parties' not falling within the forbidden degrees of kindred; Justinian was even more concerned with these than Gaius had been, although there were no significant changes in the rules in the 400 years between them. The one exception was the permission recorded by Gaius (G.1.62) for a man to marry his brother's daughter. This was a special case, however, arising from the peculiar situation of the emperor Claudius and his wife Messalina, and general advantage does not ever seem to have been taken of this licence.[18] Justinian expresses considerable horror at the concept of incest – his language is much more emotional than that of Gaius – but we know from both *Digest* and *Codex* that incestuous marriages, if inadvertent or confessed, were sometimes not dealt with as criminal at all, provided the parties ceased their union. Justinian adheres to the traditional rules, not the more restrictive ones of the canon law, in allowing marriage between first cousins (J.1.10.4) and also between step-siblings, the children of a husband by a former marriage and of a wife by her former marriage (J.1.10.8). He did not, however, permit marriage with the child of a former spouse by another marriage (J.1.10.9), even though there was no blood relationship, citing the opinion of the noted jurist Julian, who had lived in the first half of the second century AD, a generation or so before Gaius. (The continuing influence of the jurists of the early Empire, at least as creators of a

tradition, is here again apparent.) A slave's blood relationships, although not in theory known to the law, also operated once he or she was freed and so able to marry; so too, although not specifically mentioned, did the blood relationships of a bastard. Adoptive ties barred the capacity to marry, but if they were not between ascendant and descendant but between collaterals, marriage became possible if the adoptive tie were dissolved (J.1.10.2 - 3). There is again an incidental illustration of paternal power (J.1.10.2) in the reference to a man who wishes to adopt his son-in-law or daughter-in-law.

The other restrictions mentioned in J.1.10.11 include the bars to marriage between a guardian (and later a supervisor also) and his ward, between a provincial official and a woman native to the province in which he was serving, and other cases where an abuse of power was possible. It seems probable that during much of the Principate, legionaries (at least those below the centurionate) could not marry at all during their term of service.[19] In the Christian Empire, spiritual affinity, the relationship between godparent and godchild and its ramifications, became a bar to marriage.

(b) Marriage with marital subordination
(G.1.108 - 113, 115b, 136 - 137a, 148)

Although for Gaius 'free' marriage was normal, he also records the form involving marital subordination which had been the dominant, though never it seems the sole, form in earlier times. The requirements of age, consent, and capacity were the same as for free marriage, but the effects as we have seen (section II(d) above) were very different. Marital subordination by usage was possibly the most common form in early law because it was a state into which a woman fell by default. It seems to have largely disappeared towards the end of the Republic, perhaps by a reinterpretation which required there to be an intention to enter it, as otherwise 'sheer disuse' would not have had this effect.[20] Marital subordination by sharing of bread was a religious ceremony which appears to have been limited to patricians. It had faded from use in the Republic but was revived in the very early Empire, when it took effect only as far as was necessary for religious purposes, as we learn from both Gaius (G.1.136) and Tacitus (*Ann.* 4.16). Marital subordination by a contrived sale would seem to have lingered on, but the use of such a contrived sale seems to have been more common for the purpose of a formal trust, to change a guardian, etc.

The existence of free marriage even in early times, as attested by the *Twelve Tables*, is enough to prove that it was not a 'liberated' view of women that produced it, although it was certainly more compatible with the greater freedom before the law enjoyed by women in the Principate. So long as there were restrictions on women's making wills, it was in the interests of a woman's agnatic family that her acquisitions should be her own so that, on her death, they would go to that family. Marriage with marital subordination meant that any legacy, say, left to her would go to her husband if he were still alive or, if he were dead, ultimately to her new agnatic family, her husband's.

IV. Emergence from Family Authority
(J.1.12, 16)

(a) By the death of the head of the family
(J.1.12 pr - 3, 5)

Someone emerging from paternal power (or from marital subordination) as a result of the death of the man to whose authority he or she was directly subject suffered no status-loss (see section (b) below).[21] On the death of a grandfather, only his children, not the children of his sons, became independent; the children of his sons now had their biological father as head of the family (J.1.12 pr; G.1.127). 'Father' in what follows is taken to mean the senior ascendant male agnate, the *paterfamilias*, whatever his biological status. On the death of a father, all his children, female as well as male, became independent of power. Throughout much of Roman law women were, as we shall see, under guardianship, subject to restrictions not imposed on adult males, but nevertheless they too came out of power and acquired independent status. Subjection to guardianship was only the third level of classification of persons in Roman law (see section V below).

'Death' usually meant ordinary, physical death, but the same effect was consequent upon civil death, when a man lost his citizenship or even his liberty as well (see section (b) below). Justinian follows Gaius (G.1.128) here in describing the consequences of a criminal conviction which brought this about (J.1.12.3). In contrast, Justinian points out (J.1.12.4) that an act of imperial grace could release from paternal power a man whom he had appointed to the patriciate, the highest rank in the later Empire. This exercise of power was not known to the emperors of the Principate. In pagan Rome, however, priests of Jupiter and Vestal Virgins were on their selection automatically exempt from paternal power (G.1.130, 145).

A more legally interesting form of civil death took place when a man lost his liberty and citizenship through capture by some external enemy. As we saw (section I(a) above), it was generally recognised that such capture made a man a slave, putting him entirely in the hands of his captors. However, honourable return was possible from captivity as a prisoner of war. Such honourable return gave a right of rehabilitation (*postliminium*: J.1.12.5; G.1.129). The rehabilitated man could recover his rights, such as ownership, and also paternal power (and guardianship: G.1.187). Relationships, however, were not so easily recovered as rights. Thus in classical law he did not resume his marriage but needed to consent to it again, as did his former wife; in Justinian's law, under Christian influence, a marriage survived captivity, although it might end on a presumption of death. But paternal power was a right, and thus could be recovered if the father returned. If he died in captivity, he died as a slave and, technically, his will should have been void, but from expediency and because such enslavement was not in itself dishonourable, the fiction was conceded that he had died at the moment of capture and his will had thus become operative at that point. But all this might involve considerable waiting to see if a man did or did not return from captivity; during the period of suspense, children of full age must, one assumes, have acted as though independent.

(b) Status-loss
(J.1.16)

Status-loss (*capitis deminutio*), perhaps more helpfully called status change, could involve loss of liberty, loss of citizenship, or simply loss of agnatic family.[22] The first two, in the first and second (or intermediate) degrees, are straightforward enough and always a bad thing; we have noticed them above as civil death. Status-loss in the third degree was not necessarily disadvantageous; it meant a change in one's agnatic family, whether by adrogation or adoption (only the fuller form in Justinian's day), emancipation or (in the law of Gaius' time) entry into marital subordination whether for purposes of marriage or of a formal trust. It represented the loss of an individual to his or her original agnatic family, but could mean an improvement in status, as when a plebeian was adopted by a patrician. It involved complications in the pursuit of existing obligations, since the person who had undergone status-loss was in many ways a new person in the eyes of the law, but procedural solutions were found to deal with this.

(c) By emancipation
(J.1.12.6 - 10)

Emancipation was a deliberate act of release from paternal power, involving status-loss. In Gaius' time it was a ritual needing three mancipations for a son and one mancipation for a daughter or grandchild (G.1.132).[23] Justinian refers to the old procedure but proudly announces that he had 'put an end to the play-acting' (J.1.12.6). As with adoption, all that was now required was a declaration to the competent court. Again, the reality of paternal power is illustrated by the ability of the head of the family to release anyone in his power, 'to discharge the son from his authority but keep the grandchild or, vice versa, to discharge the grandchild and keep the son' (J.1.12.7). Moreover, a father could emancipate his son when his daughter-in-law was already pregnant, in this way still keeping the baby under his authority, regardless of the wishes of its parents, because a legitimate child took its status, including agnatic status, from the moment of its conception. On the other hand, there was no real way by which someone could compel his father to emancipate him. This is in contrast to a woman in a marriage with marital subordination who wanted a divorce; while she could only be released from such subordination by a mancipation, she could, after sending notice of divorce, compel her husband to release her (G.1.137, 137a). Further, a woman who had made a contrived sale of herself for the purpose of a formal trust could compel the man to whom she was subordinate to remancipate her to whomever she wished.

Justinian also points out that an emancipating father retained residual rights over the property of his emancipated child, just as a patron did over the property of a freedman. This property might, of course, consist of what had been the child's (or slave's) personal fund; it would also include what was subsequently acquired. The main element of these residual rights was the right of intestate succession where there were no children born as immediate heirs to an emancipated son, but they also included the right of guardianship where a child was still under puberty, or over a female as long as the perpetual guardianship of women was applicable. A father could in this way ensure that his daughter's property came back to the agnatic family by refusing to allow her to make a will leaving it elsewhere, just as a patron could control his freedwoman. Justinian also classes with emancipation the giving of a child in adoption to the grandfather or great-grandfather of the person to be adopted, that is, the kind of adoption that had full effect in the old way (J.1.12.8). Gaius also refers to adoption in this context (G.1.134); he then goes on to comment on release from the other two forms of family authority known to the classical law, release from marital subordination (G.1.136 - 137a) and from bondage (G.1.138 - 141).

V. Guardianship and Supervision

(J.1.13 - 26; G.1.144 - 145, 148 - 154, 157, 168 - 171, 173 - 181, 183, 190 - 195c)

The third level of classification for persons was between those who were subject to guardianship (*tutela*) or supervision (*cura*) and those who were not. Only those independent of power could have a guardian or supervisor, since the authority of a head of a family made them unnecessary. Guardianship applied to those under puberty, and in Justinian's law to them only; in classical law guardianship could also apply to adult women (as we shall see in section (f) below). Supervision applied to those between puberty and the age at which they were (in the Empire) held fully capable of looking after their own affairs, usually but not invariably fixed at 25. The supervision of minors had become a regular appointment only under the emperor Marcus Aurelius (AD 161-80); unlike guardianship it was never technically obligatory. Supervision of the insane and of others of any age who were not competent is only mentioned in passing (J.1.23.3 - 4); this institution went back to the time of the *Twelve Tables*, when such care was a family responsibility. Indeed, originally guardianship was created to preserve the interests of the general agnatic family rather than those of the individual ward. This explains why guardianship was over the property not the person of the ward, and is illustrated by the survival into classical law of the guardianship of adult women (G.1.189 - 192). It also explains the reasoning behind a patron's right to be guardian to his freedmen or freedwomen (J.1.17) in that it was someone's heirs on intestacy who had the right by state law. However, although these traces remained, by Gaius' time it had already long been a protective relationship, imposing a duty, even a burden, on the guardian: 'It is accepted that the office of guardian or supervisor is a public duty.' (J.1.25 pr).

(a) Creation of guardianship
(J.1.13.3 - 5; 1.14 - 15; 1.17.20)

The classification of kinds of guardianship seems to have concerned the jurists rather more than we might think necessary (G.1.188); of those whose views were cited by Gaius here, Q. Mucius Scaevola lived in the late Republic, as did Servius Sulpicius, while Labeo was a contemporary of the emperor Augustus. Guardians were of three main kinds: testamentary, statutory, and appointed. In some circumstances we find guardians called fiduciary and assignatory.

A *testamentary guardian* (J.1.13.3 - 5; 1.14; G.1.144 - 154) was one appointed in his will by the head of a family for those persons in his power who would become independent on his death; a mother or a brother could not appoint a guardian, nor could one be appointed for a grandchild who would fall into the power of his real father on the testator's death. Justinian points out that it was possible in a will to free one's own slave and appoint him guardian; he lays down that in that case, even if the explicit grant of freedom was omitted, it should nevertheless take effect. If the slave belonged to someone else, it was necessary to make the condition 'when he be free', for a man could not exercise guardianship while a slave, and one could not deprive another man of his property, i.e. of his slave. In Justinian's law, moreover, it was not possible for someone to exercise guardianship until he himself was of full age at 25, whereas it was technically possible from puberty in classical law; this applied to all kinds of guardian.

Statutory guardians (J.1.15, 17 - 18; G.1.155 - 158, 163 - 166, 167, 172) were those who acted, in default of a testamentary appointment, in accordance with the *Twelve Tables* and the old state law. For freeborn children the statutory guardian was the person or persons in the degree of nearest agnate (for example two paternal uncles, as nearer relations than the sons of another, deceased, paternal uncle). It is at this point in their exposition that both Gaius (G.1.156) and Justinian (J.1.15.1) describe the agnatic family as those persons traced only through the male line, but (as with our system of surnames) including women in the family to which they were born. Women were thus not excluded from agnatic relationships, but they could not transmit them. Since the agnatic family was, in a sense, artificial compared with the cognatic family (which was created by simple blood relationship), it could be destroyed by status-loss. 'While the logic of state law can destroy rights founded on the state law, it cannot so easily affect rights founded on the law of nature.' (J.1.15.3).

Other statutory guardians were the father who had emancipated someone in his power, and the patron who had manumitted a slave. (Once the emperor Claudius had abolished the statutory guardianship of agnates for an adult woman, these were the only guardians who had real control of her actions.) Since emancipation or manumission of those under puberty cannot have been usual, these kinds of statutory guardian will not have been common for children. Patronal rights differed, however, depending on whether it had been a master or mistress who freed the slave. A woman had rights of succession to her freedman, but she could not exercise the office of guardian because in earlier law she herself would have been in guardianship, and even after she ceased to be in guardianship she could not exercise a public office (J.1.17; G.1.195, 195c).

Fiduciary guardianship sprang from the statutory guardianship of an emancipating father (J.1.19); it was the only way an agnate could have statutory powers of any sort over a collateral, but even so it was not classed as 'statutory', as that of a patron's male descendants was. Gaius also classes as fiduciary the right to guardianship which followed the manumission by some man other than the original head of the family of any free person, after a mancipation for change of status (G.1.166a); this could also apply to a freedwoman (G.1.195a). Even in classical law such a guardianship could only be held by someone who was of age.

Appointed guardians (J.1.20; G.1.184 - 187, 195, 195b - 195c) were those appointed by a magistrate under statute in default of a testamentary or statutory guardian. Gaius also distinguishes here (G.1.184) the praetorian guardian who, when actions in the law (*legis actiones*) had been in use, was appointed by the urban praetor when some matter was in dispute between the ordinary guardian and his ward. Justinian simplified the procedures for appointing such a guardian, particularly where the estate of the child in question was small. There might be need for such an appointment when a testamentary guardian had been appointed under a condition which had not yet been fulfilled or when the ordinary guardian was taken captive, but in Justinian's time the relevant magistrate always held an inquiry into the circumstances.

'Assignatory' guardians are also mentioned by Gaius (G.1.168 - 172), but in practice they were relevant only to the guardianship of adult women since, as he explains, a guardian was not allowed to assign the guardianship of a child to somebody else 'because this [guardianship] is not seen as burdensome, finishing as it does with puberty' (G.1.168).

(b) Endorsement by a guardian
(J.1.21)

In the case of guardianship of a ward under puberty, the task was one of supplementing what was defective in the ward. The guardian's duty was to administer his ward's estate and, if the ward was old enough, also to endorse – give *auctoritas* to – the ward's own transactions. The ward was the principal, the one who ought to make decisions and put them into effect (if he was old enough to understand what he was doing), although because of his youth the intervention of a guardian was usually, but not always, necessary.[24] Someone under puberty could be the beneficiary of a legal transaction, but he (or she) could not bind himself. This is why the guardian's endorsement was needed for such things as the acceptance of an inheritance (J.1.21.1). Even where it was undoubted that the estate as a whole was profitable, there would inevitably be a few unsettled debts, and these a ward could not validly undertake. The endorsement of a guardian, as opposed to that of a supervisor, had to be given on the spot. Naturally enough, a guardian could not give an endorsement in a matter which affected himself. Such a conflict could easily arise if a father appointed a neighbour as guardian or if the guardian was a relation and was made, say, joint heir with his ward to some other member of the family. In such cases Justinian required that a supervisor be appointed, whereas in the law of Gaius' time an alternative guardian was appointed temporarily (G.1.184).

Justinian states that 'since guardians are provided for people, not for plans or property, it is impossible for an appointment to be made for specific property or a specific venture.' (J.1.14.4). This was not precisely true, even for his law, in that the guardian did not, as guardian, have the care of the ward's person; this seems in practice to have been left to his mother, if that were possible. The guardian of a ward under puberty was to provide funds for the ward's support and education in a style suited to his station, and also to hand over an improved rather than a diminished estate when the ward reached majority. In classical law, when there were several guardians, specific spheres of responsibility could be divided among them, or all the guardians could act jointly; it was also possible for them to choose one of their number to act on behalf of all, and this seems to have been Justinian's preferred solution (J.1.24.1).

(c) Excuses
(J.1.25)

Justinian devotes a long title to the excuses which could be offered by a potential guardian to justify avoidance of the duty. The subject must have been a frequent cause of friction, since considerable space is also given to it in the *Digest*.[25] (Again, Justinian thought it worth while to cite the view of Papinian on those excused through being away on state business, even though only to disagree with it (J.1.25.2). Papinian, however, was one of the jurists given authoritative status in pleadings before the courts of the later Empire by the Law of Citations of AD 426; see Chapter 1, III(e) above; mention of his view is therefore unsurprising.) Justinian's law also differed from that of the classical period in that for him supervisors had exactly the same range of excuses as guardians. In early classical law, as we shall see in section (e) below, since supervision was mainly needed only for specific transactions, it was not a burden which men usually sought to avoid.

For Gaius, guardianship still retained enough of its original purpose of protecting the family interests to make it possible for a child to be technically a guardian, even though he was himself under guardianship and could not exercise the office until he came of age at puberty (G.1.177 - 179). Youth might be pleaded as an excuse, but circumstances would dictate whether it was sufficient. Under Justinian, however, there was a flat prohibition on anyone under 25 even aspiring to act as guardian or supervisor (J.1.25.13). In his time supervision normally lasted until the minor was 25 and so minors were barred until themselves released from restriction. There was also a ban on soldiers' holding these offices (J.1.25.14), perhaps because they were liable to be called away on military duty, but more likely because in other contexts we find them classed with women and rustics as ignorant of affairs. Even when adult women ceased to be in guardianship, they could not themselves act as guardians because a guardian's endorsement in a sense represented paternal power and, further, guardianship was a male office. Nevertheless, from AD 390 (C.5.35.2) in special circumstances a mother or grandmother might act as guardian and, if she was widowed and undertook not to remarry, Justinian allowed such guardianship as a regular thing (Nov.118.5). For certain other classes of person, professors and doctors, exemption from the burden of guardianship was a recognised privilege (J.1.25.15).

(d) Guardians' liability and the termination of guardianship (J.1.22, 24, 26)

The guardianship of children ended when they reached puberty (J.1.22; G.1.196), originally, we are told, a matter of fact (at least for males, but surely *a fortiori* for females), but later fixed at 14 for boys and 12 for girls. In classical law this marked their coming of age. A girl would still have a guardian, but in a very different sense (as we shall see in section (f) below); where practical realities meant that some sort of endorsement was needed for the transactions of a boy over 14, a supervisor would be appointed (as we shall see in section (e) below). Even in Justinian's law, where the powers of guardian and supervisor were largely assimilated, there was still a moment of taking stock at 14 or 12, when the guardian could be called to account for his administration and his endorsements (J.1.20.6 - 7; 1.23.5 - 6; 1.24.2; G.1.191). Guardianship also ended if the child underwent any degree of status-loss, although for the guardian's own status-loss to affect the relationship it had to be in the first or second degree (J.1.22.4). This was because guardianship was a public office which could be held by someone still in his father's power; a change in the guardian's agnatic family would not affect his capacity to be guardian. An independent child, on the other hand, who changed his agnatic family, must always have been adopted (more properly, adrogated); and once in paternal power, he could not have a guardian. Emancipation, noxal surrender into bondage, and contrived sale of herself by a girl were all legally or factually impossible for an independent child, and other cases of status-loss were unlikely for a child (J.1.22.1).

The calling of the guardian to account was made easier by the fact that many guardians had to give security (J.1.24; G.1.199 - 200); only those appointed by will or by a magistrate after inquiry were exempt. Although Justinian states that poverty was not in itself a disqualification from guardianship (J.1.26.13), it must often have operated that way since the security required of the guardian had to be compatible with the ward's estate. A guardian was liable not only for fraud but also for negligence; the requirements of good faith made him liable for a high standard of care. He was required to invest his ward's capital in land, if possible, to dispose of fruits, etc. at the peak of their value, to discharge debts when they fell due so as to avoid being liable for interest, and to pursue debts owed to the ward's estate. Limits were put on his powers of alienation, although the competent court might waive these for good cause. He must give his endorsement to his ward's transactions with the care required of the reasonable man, and he must render proper accounts. The action on guardianship was available at its termination to the former ward, who also had a policy action on guardianship if the guardian had failed to act at all; the guardian might have a counter-action to recover his expenses.

If it was felt necessary to bring the guardianship to a premature end, a charge of untrustworthiness could be brought before the relevant court by anyone having an interest, including, for example, the mother, but not the ward himself, although after he reached puberty he could bring such a charge against a supervisor. Such a charge, which was within the ambit of the criminal law, led to the guardian's immediate suspension; Justinian cites Papinian again to confirm this point (J.1.26.7). Justinian also cites the jurist Julian to support the view that this charge could be brought on general grounds of character or behaviour elsewhere, even before the guardian began to act (J.1.26.5). The death of the suspect guardian brought any trial to an end (J.1.26.8), although his heir would be liable for any unjustified enrichment of the guardian's estate. Where the guardian was found to be dishonest, he had to be removed from office; security for future good conduct was only acceptable where the guardian had been no worse than negligent (J.1.26.12). Where the guardian had acted fraudulently, condemnation involved him in infamy, although respect for the special position of a patron – as former owner with total power – might enable him to avoid this (J.1.26.2).

(e) Supervision of minors
(J.1.23)

A page is missing from Gaius' account of supervision, but even so it seems clear that he was not so concerned with the institution as was Justinian. The relatively late emergence (around 200 BC) of supervisors for young persons explains why they were always appointed by a magistrate, even when in confirmation of the will of a deceased head of a family. Minors (the term comes from the Latin phrase 'minor quam viginti quinque annis', 'less than 25 years old') gave their consent, and therefore could withhold it, to the appointment of their supervisor (J.1.23.2). Originally, indeed, a minor himself applied for a supervisor for a specific transaction. The emperor Marcus Aurelius seems to have generalised the institution so that a supervisor, once appointed, would continue to hold office without limit other than the minor's attaining majority. In post-classical law a supervisor was appointed automatically for any minor who came to the notice of the court. The supervisor could give his assent to the minor's acts retrospectively or by letter; this assent came to be required, as with a ward, for all cases where the minor's legal position could become worse, however technically. In classical law the minor might entrust the administration of his affairs to his supervisor as a general agent; in post-classical law, administration came to be a function of a supervisor as it was of a guardian. This explains why the rules on excuses to avoid the office, on the giving of security for its proper conduct, and on the bringing of a charge of untrustworthiness, came to be identical for guardians and supervisors; Justinian's harmonisation of the two offices was virtually complete.[26]

(f) Guardianship of women
(G.1.144 - 145, 148 - 154, 157, 168 - 171, 173 - 181, 183, 190 - 195c)

The guardianship of women was undoubtedly in the interests of the agnatic family; that was why the head of a family could in his will appoint a guardian even for a married daughter (G.1.144), and why a husband as such had no claim to act as guardian for his wife. This purpose was rendered largely nugatory by the Claudian Act which abolished agnatic guardianship for women (G.1.157); after this, the only statutory guardians left for adult women were the emancipating father and the patron. The major remaining aspect of the predominance of the agnatic family over the individual woman was removed by the emperor Hadrian's legislation, which allowed women to make a will without first having to undergo status-loss (G.1.115a; 2.112); thus it gave this capacity to women when they became independent by the death of their father. This is why Gaius writes that there seemed to be no very convincing reason for the guardianship of adult women (G.1.190). By his time, the only guardians who could not readily be changed were the remaining statutory guardians, and even they had no power to administer or to compel the woman to act; all they could do was refuse endorsement. This was why the guardians of adult women could not be held liable for their guardianship, unlike the guardians of children (G.1.191); they had no power to do anything for which they could be held accountable.

Not only was the guardianship of women, where it existed, emptied of almost all real content, but for many women it no longer applied. Under the Julian and Papian-Poppaean Acts, by which Augustus regulated marriages in 18 BC and AD 9,[27] the privilege of children was granted to free-born parents who had had three legitimate children. For women this meant release from guardianship; freedwomen attained this privilege by having four children. While it seems likely that many freedwomen would be unable to take advantage of this privilege if most of their child-bearing years were spent in slavery, yet it must have been common among free-born women who lived so long. Guardianship of women may have disappeared as an institution before the end of the 3rd century; Justinian makes no mention of it.

Select Bibliography

Slavery

Buckland, W.W., *The Roman Law of Slavery* (Cambridge, 1908).
Fear, A.T., '*Cives Latini, servi publici*, and the *lex Irnitana*', *RIDA* (3rd ser.) 37 (1990) p. 149.
Finley, M.I. (ed), *Slavery in Classical Antiquity* (Cambridge, 1960).
Robinson, O.F., 'Slaves and the Criminal Law', *ZSS* (rom. Abt.) 98 (1981) p. 213.
Treggiari, S., *Roman Freedmen during the Late Republic* (Oxford, 1969).
Watson, A., *Roman Slave Law* (Baltimore, 1987).

Family Authority and Emergence from Family Authority

Crook, J., '*Patria potestas*', *CQ* (new ser.) 17 (1967) p. 113.
Rawson, B. (ed), *The Family in Ancient Rome* (London, 1992).
Thomas, J.A.C., 'Some Notes on *adrogatio per rescriptum principis*', *RIDA* (3rd ser.) 14 (1967) p. 413.
Watson, A., *The Law of Persons in the Later Roman Republic* (Oxford, 1967) ch. 8.

Marriage

Corbett, P.E., *The Roman Law of Marriage* (Oxford, 1930).
Looper-Friedman, S.E., 'The decline of *manus* marriage', *TvR* 55 (1987) p. 281.

2. Persons

Pomeroy, S.B., 'The Relationship of the Married Woman to her Blood Relatives at Rome', *Ancient Society* 7 (1976) p. 215.
Rawson, B., 'Roman Concubinage and other de facto Marriages', *Transactions of the American Philological Association* 104 (1974) p. 279.
——— (ed), *Marriage, Divorce and Children in Ancient Rome* (Oxford, 1991).
Treggiari, S., *Roman Marriage* (Oxford, 1991).

Guardianship and Supervision

Gardner, J.F., *Women in Roman Law and Society* (London, 1986) pp. 5-29.
Jolowicz, H.F., 'The Wicked Guardian', *JRS* 37 (1947) p. 82.
MacCormack, G., 'The Liability of the Tutor in Classical Roman Law', *IJ* (new ser.) 5 (1970) p. 369.
Robinson, O.F., 'The Status of Women in Roman Private Law', *JR* (1987) p. 143.

Notes

1. There is no discussion, in either book, of citizenship, on which see A.N. Sherwin-White, *The Roman Citizenship* 2nd ed. (Oxford, 1973); cf. J.F. Gardner, *Being a Roman Citizen* (London, 1993).

2. See P.R.C. Weaver, *Familia Caesaris* (Cambridge, 1972) pp. 162-9.

3. 17 BC and AD 19 are the dates suggested; the latter rests solely on the reference by Justinian (and by him alone) to the *lex* as *Junia Norbana*. An Augustan date seems intrinsically more likely; 17 BC is suggested by H.F. Jolowicz and B. Nicholas, *Historical Introduction to the Study of Roman Law* 3rd ed. (Cambridge, 1972) pp. 136, 345. Gaius appears to make the Aelian-Sentian Act of AD 4 refer to Junian Latins (G.1.29).

4. On informal manumission see A.J.B. Sirks, 'Informal Manumission and the *lex Iunia*', *RIDA* (3rd ser.) 28 (1981) p. 247; id., 'The *lex Iunia* and the Effects of Informal Manumission', *RIDA* (3rd ser.) 30 (1983) p. 211.

5. See K.M.T. Atkinson, 'The Purpose of the Manumission Laws of Augustus', *IJ* (new ser.) 1 (1966) p. 356.

6. P.A. Brunt, 'Free Labour and Public Works at Rome', *JRS* 70 (1980) p. 81; M.K. and R.L. Thornton, 'Manpower Needs for the Public Works Program of the Julio-Claudian Emperors', *Journal of Economic History* 43 (1983) p. 373.

7. The provisions of the recently discovered *lex Irnitana* cast some doubt on Gaius' categories; see G. Hansard, 'Note à propos des *leges Salpensana* et *Irnitana*: faut-il corriger l'enseignement de Gaius?', *RIDA* (3rd ser.) 34 (1987) p. 173.

8. See P. Voci, 'Storia della *patria potestas* da Costantino a Giustiniano', *SDHI* 51 (1985) p. 1.

9. T. Mayer-Maly, 'Das Notverkaufsrecht des Hausvaters', *ZSS* (rom. Abt.) 75 (1958) p. 116; M.B. Fossati-Vanzetti, 'Vendità ed esposizione degli infanti da Costantino a Giustiniano', *SDHI* 49 (1983) p. 179. For a different perspective, see A. Russi, 'I pastori e l'espozione degli infanti', *Mélanges d'archéologie et d'histoire de l'École Française de Rome* 98 (1986) p. 855; cf. J. Boswell, *The Kindness of Strangers* (New York, 1988).

10. D. Daube, 'Actions between *paterfamilias* and *filiusfamilias* with *peculium castrense*', *Studi in memoria di E. Albertario* I (Milan, 1953) p. 433.

11. See M. Kurylowicz, '*Adoptio plena* und *minus plena*', *Labeo* 25 (1979) p. 163.

12. If the castration had been voluntary, it was disgraceful, and this may be enough to explain the rule. But it seems odd.

13. The references to such a form are literary, not legal. See Kaser, *RPR* I p. 349.

14. See. O.F. Robinson, 'The Historical Background', in S.M. McLean and N. Burrows (eds), *The Legal Relevance of Gender* (Basingstoke, 1988) p. 40.

15. See K. Visky, 'Le divorce dans la législation de Justinien', *RIDA* (3rd ser.) 23 (1976) p. 239.

16. See K. Hopkins, 'The Age of Roman Girls at Marriage', *Population Studies* 18 (1965) p. 309; cf. St Augustine's *Confessions* 6.13.

17. See R. Yaron, '*Divortium inter absentes*', *TvR* 31 (1963) p. 54; id., '*De divortio varia*', *TvR* 32 (1964) p. 533.

18. O.F. Robinson, *The Criminal Law of Ancient Rome* (London, 1995) pp. 54-7.

19. See B. Campbell, 'The Marriage of Soldiers under the Empire', *JRS* 68 (1978) p. 153.

20. A. Watson, *The Law of Persons in the Later Roman Republic* (Oxford, 1967) pp. 19-23. See also J. Linderski, '*Usu, farre, coemptione*', *ZSS* (rom. Abt.) 101 (1984) p. 301.

21. E. Volterra, 'L'acquisto della *patria potestas* alla morte del *paterfamilias*', *BIDR* (3rd ser.) 18 (1976) p. 193.

22. See M. Kaser, 'Zur Geschichte der *capitis deminutio*', *IURA* 3 (1952) p. 48.

23. P.B.H. Birks, '3 x 1 = 3: an Arithmetical Solution to the Problem of Threefold Mancipation', *IURA* 40 (1989) p. 55.

24. See A. Burdese, 'Sulla capacità intellettuale degli *impuberes* in diritto classico', *Archivio Giuridico* 150 (1956) p. 10.

25. L. Pelliciani, 'D 27.1.1 *pr*.-2 e i *libri excusationum* di Modestino', *Labeo* 24 (1978) p. 37.

26. For a full treatment see G. Cervenca, 'Studi sulla *cura minorum*', *BIDR* (3rd ser.) 14 (1972) p. 235; id., *BIDR* (3rd ser.) 16 (1974) p. 139; id., *BIDR* (3rd ser.) 21 (1979) p. 41.

27. See D. Nörr, 'The Matrimonial Legislation of Augustus', *IJ* (new ser.) 16 (1981) p. 350; cf. A. Wallace-Hadrill, 'Family and Inheritance in the Augustan Marriage Laws', *Proceedings of the Cambridge Philosophical Society* (new ser.) 27 (1981) p. 58.

3. Property

D.L. Carey Miller

I. Introduction

The Roman law of property reflects similar problems of classification and labelling to those of modern law. The scope of the subject is reflected in the notion of 'things', which represents a basic division of Roman law. Central to the systems of Gaius and Justinian is the law governing persons and that concerned with things. In this wide sense the law relating to things encompasses all items having a physical existence but also a range of abstractions which amount to 'property' in the legal sense.[1] Whether one uses the term 'property' in the broad general sense of any form of asset or claim, or in the more limited sense connoting a right of property, there is room for a distinction between corporeal things, having a physical existence, and the category of intangible items designated 'incorporeal'.[2] As Barry Nicholas[3] points out, the texts of Roman law do not define the scope of the category of 'things' but 'leave its meaning to emerge from its use'.

The integrity of the Roman law of property is to be found not in the category of 'things' but in the notion of a real (or proprietary) right, whether relating to a corporeal or incorporeal thing. In a sense Roman property is not so much a self-contained branch of the law but an essential grammar with implications for contract, succession, and procedure. The central feature of this grammar is the distinction between real and personal rights which is explained in the section immediately following. Other distinctions and bases of classification (section (b) below) are also relevant, both to the scope of the notion of things and to the working of the law.

(a) Real and personal actions and rights
(J.4.6.1)

The legal means by which a right or claim was enforced is an 'action' (see Chapter 6, section I below). Roman law recognised two basic kinds of action and this distinction has important implications for property. Actions were either *in personam* or *in rem* (G.4.1 - 3; J.4.6.1). The former applied to a claim by one party against another. A claim based upon a contract or arising out of a wrongful act (a 'delict') was the usual basis of an action *in personam*. An action *in rem*, on the other hand, was one in which the claimant asserted an established entitlement to a thing, whether a corporeal item or a 'thing' in the sense of an incorporeal right. The essence of the difference is that in the case of an action *in personam* the claimant asserts a legal relationship with a particular person directed towards the conclusion that something is due to him by the person concerned. An action *in rem* is a claim in and to a specific thing on the basis that the claimant is entitled to it or to some part of it.[4]

Roman law emphasised the difference between actions *in rem* and those *in personam*, but a legal action is the assertion of a right, and the dichotomy could only be predicated upon the recognition of the difference between real and personal rights. The distinction between actions *in rem* and actions *in personam* is central to the subject of property.[5] Given the significance of the distinction between a claim, necessarily directed against a particular individual, and a real (or 'proprietary') right in a thing, by definition available on a general basis (or as is sometimes said 'against the whole world'), it follows that the identification of a real right is of considerable importance. The mere fact of possession does not give any insight into the nature of the right on the basis of which the thing in question is held. It may well be necessary to examine the circumstances under which the thing was acquired to determine the standing of the right. The Roman law of property is based upon the recognition of an ultimate right of ownership superior to all other possible rights in a given thing. The various ways in which ownership could be acquired are an important part of the law of property.

A distinction associated with the one between real and personal rights is that between contract and conveyance. The label 'contract' identifies an act which gives, or the prevailing state of, a mere claim by one party against another – the basis for an action *in personam* or the condition of a right *in personam*. In the context of property 'contract' connotes a claim to a thing but, importantly, not an established right in the thing. 'Conveyance' identifies the legal act which causes the right of property in the thing concerned to pass from one party to the other. The legal act of conveyance gives the acquiring party an established right in the thing. In the common case of sale, upon the conclusion of a contract between seller and buyer, reciprocal obligations – each involving a particular claim (or right) and a correlative particular obligation (or debt) – come into being. The act of conveyance is a separate process, also involving the agreement of the parties, on the basis of which the buyer's claim and the seller's obligation are given effect by the transmission of the right of property in the thing from the seller to the buyer.

(b) Classifications and distinctions
(J.2.1, 2 pr - 2)

Corporeal and incorporeal things. Roman law recognised the distinction, already referred to, between corporeal and incorporeal things or property. There is no difficulty with the category of corporeal property which extends to all tangible things (J.2.2.1; G.2.13). The category of incorporeal property should be restricted to rights which are real or proprietary (in the sense explained above) because a merely personal claim, although clearly a right, is not property in the sense of being an entitlement available against the whole world. But Gaius (G.2.14) refers to the wide category of incorporeals including 'obligations however contracted'. A contractual right is incorporeal, but not being available against all it is not property in the sense that a right of servitude is (see also J.2.2.2: '[incorporeal things] consist of legal rights'). In the loose sense of property as an asset or something of value, the category of incorporeals may include a claim available only against a particular individual. If, however, one is thinking of property in the technical sense of a proprietary right – a vested interest available against all – then a contractual (or delictual) obligation is not property.

Res mobiles and res immobiles. Roman law recognised the obvious actual difference between land and the buildings and other structures which formed part of it (*res immobiles*), and moveable things (*res mobiles*). In common with the approach of most legal systems, Roman law made land the subject of special treatment by the rules of property.[6] But rights in land did not dominate private law as they did in subsequent European feudal systems.[7] It would probably be accurate to say that in Roman law the principles of the law of property apply equally to land and to moveable things. The differences are matters of detail tending to reflect considerations of policy.

Res mancipi and res nec mancipi. Through much of the development of Roman law the transmission of rights in property – conveyancing – proceeded on the basis of a distinction between two classes of property. *Res mancipi* included corporeals and incorporeals which were taken to be of such importance that a special form of conveyance was prescribed. The category included land, real rights in land, slaves, and beasts of draught and burden. *Res nec mancipi* were all other things, in respect of which property could be passed by an informal act of conveyance. Although this distinction was based upon considerations of policy rather than any more fundamental factors, it had far-reaching implications and, for practical purposes, it was a consideration overriding anything deriving from the distinction between moveable and immoveable things.

Other bases of classification. The widest sense of 'things' includes all items having an identifiable independent existence, whether as physical or abstract entities. But in Roman as in modern law, certain classes of property had to be distinguished and were subject to special rules or limitations (J.2.1). A first distinction was that things are either within the private domain or excluded from it (G.2.1). A category of things subject to divine right – identified as either *res sacrae* or *res religiosae* – were not open to ownership (J.2.1.7 - 9; G.2.3 - 9). Things potentially subject to human right could be either public (*res publicae*) or private (*res privatae*), the latter being open to individual ownership but the former not (G.2.10 -

11). Things open to ownership by individuals – the important category for the private law of property – could be either owned or unowned. An unowned thing (identified as *res nullius*) was open to acquisition by anyone capable of exercising the necessary intention to acquire it (see section III(g) below).

II. Ownership and Possession

(a) Ownership

The concept of ownership (*dominium*) as the legal relationship between a person and a thing giving the greatest possible accumulation of rights was well established in Roman law[8] and it has been passed on to many modern systems. What is significant about the Roman notion is that ownership is a distinct paramount right rather than a mere label attaching to the most compelling of two or more competing claims to a thing. In this latter sense ownership is essentially a condition or state of superiority. Roman ownership is more than the according of a right on the basis of priority over actual or hypothetical competing rights. Rather, it is the established ultimate right which could be obtained through some appropriate process of legal acquisition. Moreover, in Roman law, the title of ownership carries with it the strongest proprietary remedy – the vindicatory action – in principle available against all other potential claimants. Significantly, this remedy is distinct from the lesser remedies open to the holders of lesser rights (see Chapter 6, section I below).

The privileged case of Roman ownership – implying 'a Roman owner of a Roman thing acquired by Roman process'[9] – was very much concerned with an ultimate right. But it was recognised as a matter of natural law that ownership was a corporate concept, an ultimate right encompassing identifiable constituent rights. The breakdown into constituent rights is important in the recognition of the lesser rights deriving from ownership and of the potentially independent right of possession.

The essence of the notion of an ultimate right is the power to dispose of the thing in question: the control of the right to convey the ultimate right itself. Provided this right is retained, ownership is retained. Being the total of possible rights, Roman ownership in its complete form includes the rights to the use and fruits of a thing, the rights which follow as a benefit from 'having' the thing. But an owner could part with these rights and remain owner, provided control over the right of disposal was retained. The composite nature of the right of ownership is reflected in the description *usus*, *fructus*, *abusus* (the right to use, the right to fruits, and the right of disposal),[10] though this does not identify the right of disposal (*abusus*) as the essential element.

The actual use and actual access to the fruits are only available to a party in physical control of a thing. Ownership, encompassing the rights to use and to fruits, must necessarily include the right to physical control – the right of possession. But it should be noted that the word 'possession' has various meanings; this will be more fully considered below.

Ownership can only be present where the abstract condition of legal title exists. Although there may be a right to possession as an abstract condition, the essence of possession, as an independent concept, is the actual holding of a thing. Ownership as an abstract condition can, in principle, come into being without any tangible link between the owner and the thing (although legal systems frequently require a link). Possession, on the other hand, being fundamentally a matter of physical state, can only come into being on the basis of some form of actual holding or control. This distinction between the abstract notion of ownership and the physical condition of possession is of some importance in the Roman law of property. Of particular importance is the distinction between the right of ownership as a provable condition, and the particular form of possession involving the holding of a thing on the basis of a belief in a right of ownership.

(b) Possession

Unlike ownership, possession is something which exists and can be defined without resort to legal criteria. Its ordinary meaning is the holding of a thing, the condition of the actual holding of a thing being designated 'natural possession'. On this basis someone who has a thing within his actual physical control may be said to 'be in possession'. Possession becomes a matter of legal recognition because of its importance in any regulated system of property rights. A legal system which sought to control property rights without regard to the fact of possession would be unrealistic and probably unworkable.

Property systems organised on the basis of the notion of ownership must regulate the relationship between ownership and possession because, clearly, someone other than the owner may be in possession. This regulation necessarily entails the recognition of different forms of possession. The circumstances under which a thing is held vary, and the law must take account of this in determining the standing to be accorded a particular form of possession. In the law's recognition of various forms of possession some forms are protected because of their importance within the property system. As a matter of primary classification Roman law distinguishes the instances of possession as a mere physical holding and possession as a protected right. Protected possession is something more than a mere physical holding because it involves a legally protected right. In so far as a deprived possessor has the right to recover his lost possession, one can identify a 'right to possession' prevailing even though actual (or natural) possession has been lost. Moreover, a party enjoying a right to possession in this sense may part with the thing on some temporary basis. In these circumstances, provided the residual right to possession remains extant, the legal possessor may recover the thing upon the expiry of the transient right of the party holding the thing.

The important form of possession in Roman law was that in which the possessor held the thing as owner. Put another way, one might say that it was the case of a thing being held on a basis not inconsistent with the right of ownership. Someone who had natural possession on the basis of a contract with the owner, or through a limited right of property (e.g. usufruct; see section IV(b) below) which acknowledged the ownership of another, did not have possession in this sense. One who did possess as owner – typically a party who had acquired the thing on a basis consistent with the acquisition of a right of ownership – had a right which was legally protected and was capable of maturing into ownership (see section (c) below).

A complete and unlimited right of ownership encompassed the right to possession and the protection which attached to it. Ulpian (D.41.2.12) says that 'ownership has nothing in common with possession' because a person 'is not deemed to have renounced possession by asserting ownership'.[11] This is no more than a statement that ownership may encompass possession; where it does encompass possession, the right to possession remains an independent right, in consequence of which proprietary and possessory remedies are available in the alternative.

(c) Remedies
(J.4.6.1; 4.15)

The Roman law distinction between ownership and possession is given effect in the recognition of different legal remedies protecting the respective rights. The right of ownership is protected by a vindicatory action. The essence of this is that an owner, upon proof of his title, can recover the thing from any party not holding by way of a right available against him. The basis of this action is a right in the thing claimed and it is accordingly designated *in rem* (J.4.6.1; G.4.3). The applicable form of procedure involved the assertion of a claim to the actual thing (G.4.16).

It is easy to see that the notion of an ultimate right of ownership is only meaningful in so far as an owner can, in principle, follow up his property and recover it from anyone. But, this said, the owner's general right to regain his property must be subject to a particular right he has accorded the holder. Where B has helped himself to a book from A's library A can vindicate it, but he cannot do so where B holds the book on the basis of a loan from him. Moreover, where C bought the book from B in the honest belief that B had a right to sell it, A's right of ownership nevertheless allows him to get it back from C. C is a possessor in good faith (a *bona fide* possessor), but the protection accorded one who held even on an entirely honest belief in his right to the thing was subject to the overriding right of the true owner.

In Roman law the position of a *bona fide* possessor was of considerable importance because, in respect of many forms of property, ownership could only pass on the basis of a prescribed formal act of transfer. This meant that it was often the case that there was every justification for recognising that the *bona fide* possessor should be regarded as owner. The means by which this was achieved will be dealt with under acquisition (section III(e) below). What we are concerned with here is the protection of the right of possession itself.

As and when possession ripened into ownership the owner's right of vindication was immediately available.[12] Pending this, possession was protected in Roman law by a legal remedy known as a possessory interdict. However, this remedy was available only to one who possessed in the full legal sense and, as indicated, this meant that the protection only prevailed in the case of one who held the thing as owner. Put in terms of a criterion which could be applied without undue difficulty, in principle one who possessed on a basis inconsistent with a right of ownership did not have recourse to the remedy.[13]

An important role of the possessory interdict was to resolve the issue of the right to possession pending the resolution of a dispute as to ownership (J.4.15.4; G.4.148). The question of ownership was resolved in a vindicatory action in which the claimant would have to prove his right to the disputed thing. The onus being on the claimant meant that the defending party was in a stronger position. The party in possession was the appropriate defender and, consequently, the preliminary issue of who was entitled to possession might well be disputed.

The possessory interdict applied as an interim order that the party determined to be in possession was entitled to retain the thing pending the resolution of the question of ownership. Before Justinian's time, land and moveable things were treated differently. Where the dispute was over land, the party who was in possession – the operative time being when the application for the interdict came to court – was entitled to continue in possession provided his case satisfied certain requirements. These were that he had not obtained the land from the other party by force or deception, or on the basis of a grant. When the interdict was concerned with a moveable the interim order would be in favour of the party who had been in possession for the greater part of the preceding year, provided that he had not obtained the thing by force or deception, or by grant, from the other party (G.4.150). By Justinian's time land and moveables were dealt with in the same way, with the interdict granted in favour of the party who was in possession at the time the judicial order was made. It remained that a claimant would not succeed if he had in fact obtained the thing, from the other party, by force, deception, or on the basis of a grant (J.4.15.4a.).

Two features of the possessory remedy are of particular importance. The first is that it is not concerned with the question of ownership: its focus is solely upon the right to possession. Second, it was only available against an immediate dispossessor. This meant that while A could rely upon his right of possession to recover from B, his immediate dispossessor, he could not get the thing back from C or any other subsequent party. Where possession had passed from A's immediate dispossessor B, A's only remedy was to bring a vindicatory action. But, of course, he would only succeed if he could prove a right of ownership and, pending this, C (or any other subsequent party) was entitled to retain possession.

For further discussion on interdicts, see the discussion below in Chapter 6, section I.

III. Acquisition of Ownership

(a) Introduction
(J.2.1.11, 40)

Roman law recognised that the acquisition of ownership could be either on the basis of universally recognised precepts of natural law or through the particular provisions of a given system (J.2.1.11). Thus while it is a matter of rational conclusion that unowned property is open to acquisition by an occupier, and while delivery may be seen as a natural basis of acquisition (J.2.1.40), the manner and means by which ownership is acquired on the basis of prescription is a matter to be determined by individual legal systems.[14]

The various modes of acquisition of ownership known to Roman law reflect another difference which is important to the analysis of acquisition. This is the distinction between original and derivative acquisition. Acquisition is original where the acquirer obtains a title *de novo* (on an original basis) as in the case of someone taking to himself a thing not previously owned and so open to acquisition by the first taker.[15] Acquisition is derivative where a party, already owner, transmits his right to another who receives and so becomes owner. Clearly, this is the most usual and most important basis of the acquisition of ownership in any developed economic system.

The form which derivative acquisition takes varies between different legal systems. While there is a universal notion of derivative acquisition in a general sense, the actual basis upon which ownership passes in a particular derivative process is subject to variation. Through much of the development of Roman law, derivative acquisition applying to the important forms of property designated *res mancipi* (see section I(b) above) was by a formal act. The involvement of transferor and transferee in a prescribed process of conveyance of the property concerned exemplifies the notion of derivative acquisition.

A feature of Roman law was the important role of usucapion (*usucapio*), a form of acquisitive prescription (see section (e) below), in correcting difficulties which arose in respect of derivative acquisition. A failure to comply with the formalities in respect of the transfer of ownership in *res mancipi* frequently left the transferee without a good title. But a transferee in this position became owner following possession for a period of time through the device of usucapion. This is an original mode of acquisition because, on a correct analysis, ownership does not derive from a predecessor but arises *de novo* following the completion of the requisite period of possession.

Because of the importance of usucapion in correcting defective derivative acquisition it is preferable to treat it after discussing the derivative modes of acquisition.[16]

(b) Derivative acquisition: mancipation
(G.1.119 - 121)

Until after Gaius' time the required mode of conveyance of the category of property designated *res mancipi* (see section I(b) above) was *mancipatio*, a fictitious sale before witnesses. The fact that *mancipatio* extended to a wide category of property, both moveable and immoveable (G.1.120), indicates an emphasis upon formality. But applying to land the purpose of *mancipatio* is unexceptionable and essentially similar to controls applied by most modern systems.

The most important purpose of *mancipatio* was to ensure compliance with the fundamental requirement of derivative acquisition: that the transferor intended to give, and the transferee intended to receive, ownership in the thing concerned. The successor of *mancipatio* was probably a written agreement reciting that the act of handing over had been complied with,[17] thus recording that the parties intended, by their act of delivery, that ownership should pass.

Gaius (G.1.119) refers to *mancipatio* as 'a sort of imaginary sale'. This is significant because it suggests that the notion of an agreement to transfer ownership – which has become accepted dogma in modern theory[18] – was recognised in Roman law.

Gaius (G.1.121) notes that the mancipation of land differs from that of other things. Moveable things had actually to be handed over and received but, for obvious reasons, there was no such requirement in respect of land. Lee[19] notes the likelihood of a clod of earth being handed over as a symbol of the land to be conveyed. Something analogous applied in the procedure applicable to a legal claim to land (G.4.17) and it was also the case that large or cumbersome moveables could be represented in court by a symbol. Lee[20] makes the point that in so far as land could be conveyed by *mancipatio* on the basis of the handing over of a symbol, 'there is no apparent reason why the same symbolism should not have been available also in the case of moveables'. But Gaius (G.1.121) is clear that moveable things 'cannot be mancipated unless they are present'. Given the potential problems associated with title to moveables one can understand the requirement of an actual handing over of the item concerned.

(c) Derivative acquisition: assignment in court (G.2.24 - 25, 28 - 31)

The second formal mode of acquisition was by a fictitious legal action in which the transferee claimed the thing and the transferor did not resist: *in iure cessio* (G.2.24). This had to take place before a magistrate and for this reason its use was less widespread than the use of *mancipatio*. As Gaius (G.2.25) puts it, there was no reason to do with greater difficulty before an official what could be done privately using friends as witnesses. What, then, was the role of *in iure cessio*?

The primary use of delivery by judicial act seems to have been in the transmission of incorporeal property. Rights not being open to actual delivery, the law must necessarily provide for a means of transmission and preferably one which promotes a measure of certainty. As Gaius (G.2.30) shows, the creation of a right of usufruct could be achieved by an owner ceding the rights to use and fruits while retaining bare ownership. A reverse cession by the usufructuary to the owner would terminate the usufruct and restore to the latter the full complement of the constituent rights of ownership. Where the cession is effected by *in iure cessio*, an element of certainty attaches to a transaction which would otherwise be lacking in external form.

In iure cessio was the only means of creation of a usufruct – or other limited real interest – by isolated act (see sections IV(b) and (c) below). Where the corporeal property concerned was transmitted and an incorporeal real right was retained, *mancipatio* could be used with the right retained excluded from the property conveyed (G.2.33).[21]

In iure cessio was the required form only in respect of *res mancipi*; limited real rights in provincial land could be created and transmitted by informal means (G.2.31). The *in iure cessio* method benefited from the security of a prescribed form of public act and, at any rate in respect of real rights in land, it would appear to be preferable to an informal act of transmission. In this respect it is closer to modern systems than the later Justinianic Roman law is.[22]

(d) Derivative acquisition: delivery (J.2.1.40 - 46)

Delivery pursuant to an antecedent agreement is the quintessential form of derivative acquisition, so much so that it is seen as a matter of natural law: 'What could be more in line with natural justice than to give effect to a man's intention to transfer something of his to another?' (J.2.1.40). But, of course, the requirement of delivery is no more than an option even though it may be seen as a natural one.[23] This said, there must be a transfer of the right of possession to give a transfer of ownership. X cannot continue to hold as owner once he has committed himself by some legal act to transfer his right of ownership to Y.

The law relevant to delivery may be considered on the following basis: (1) prerequisite conditions; (2) essential active requirements; (3) different forms; and (4) the special requirement in respect of sale.

Prerequisite conditions. The requirement that the parties have the legal capacity to act is a matter external to property but, of course, it had to be satisfied in Roman law. The precondition that ownership in the thing concerned could be transferred by delivery is more germane to property; the thing had to be open to acquisition in the sense of being susceptible to private ownership (G.2.10, 11). Within this broad category certain limitations applied.

The distinction between *res mancipi* and *res nec mancipi*, which survived in Gaius' time but lost its importance before Justinian, was a major limitation upon conveyance by delivery (see section (b) above). The other basic limitation was that the property was corporeal. Gaius (G.2.19) notes these two limitations to the scope of delivery: 'For [*res nec mancipi*] become the full property of someone else by the very act of delivery, provided that they are corporeal and so capable of delivery.'

The most important prerequisite for the transfer of ownership by delivery was that the transferor was owner of the thing concerned or vested with the right to convey on behalf of the owner. Gaius (G.2.20) notes that where I deliver a thing pursuant to a transaction envisaging the passing of ownership, I make the transferee owner 'provided that I am owner of it'. The fact that this essential is implicit in the notion of a transfer of ownership by delivery explains the indirect reference to it in Justinian's *Institutes* (J.2.1.42): 'It makes no difference whether the delivery is made by the owner or by another with his consent.'[24]

Essential active requirements. The transfer of ownership by delivery is driven by the parties' intentions. Roman law recognised that a bare act of delivery was itself meaningless, and that there had to be an intention to convey on the part of the transferor and an intention to receive on the part of the transferee: *animus transferendi et adquirendi dominii*.

Delivery came to be the primary mode of conveyance because it is a natural manifestation of the owner's intention to pass ownership (J.2.1.40). But delivery can only be effective where intention is present. The importance of intention is shown by those instances of constructive delivery where, although there is no actual handing over, there is some identifiable basis for inferring the necessary intention (e.g. J.2.1.44; see below concerning *traditio brevi manu*).

In Roman law, as in modern law, the parties' intention that ownership should pass by delivery could be inferred from the circumstances of an antecedent transaction and, clearly enough, this would be the usual position. Accordingly, one could infer the necessary intention where a thing was handed over following the conclusion of a contract of sale. The prevalence of the act of delivery following an inducing transaction led to a tendency to look for an appropriate basis from which the necessary intention could be inferred. Thus it was said that there had to be a *iusta causa*.[25]

What the requirement of *iusta causa* meant in the context of derivative acquisition in Roman law is a matter of some debate, not least because it reflects an issue faced by modern systems. Where should the law come down between, on the one hand, the requirement of a valid underlying transaction – from which the act of delivery could be seen to follow – and, on the other hand, any perceived basis sufficient to support the inference that the parties intended that ownership should pass?[26] But the question how the intention requirement was satisfied in the different stages of the development of Roman law is less important than the principle from which the debate arose, i.e., that there could be no transfer by an unmotivated act of delivery.

Because the *causa* was only evidence of the parties' intention that ownership should pass, it did not need to have an objective existence. It could be putative provided it pointed to a mutual intention that ownership should pass. Accordingly, ownership passed by delivery even where the parties were not in agreement over the *causa*, provided that the circumstances supported the inference that they intended ownership to pass. On this basis, ownership passed upon delivery where the transferor contemplated a sale and the transferee believed that he was receiving a gift.[27]

Different forms. The usual form of delivery was the actual handing over by transferor to transferee. There had to be a transfer of the right of possession – in the sense of the right to hold as owner – to give a transfer of ownership (see section II(b) above). Actual delivery was the most obvious means of achieving this. Thus where the transferor handed the thing over intending to make the transferee owner, and the latter received possession 'as owner', he immediately became owner (G.2.20: 'it immediately becomes yours').

Roman law recognised a range of alternatives to actual delivery and modern systems have been influenced by the Roman forms. In the *Institutes* (J.2.1.44) some of these alternative forms are introduced with the statement that 'sometimes the owner's mere intention, even without delivery, is enough to transfer the property in the thing'. The common requirement is some event in the relationship between the parties which can be identified as the point of transfer, on the basis that the necessary intention may be inferred from the circumstances.

It is a truism that, in departing from the requirement of an actual handing over, the alternative forms of delivery indicate that the *animus* (intention) factor has a more essential role than the *corpus* (physical act) factor (J.2.1.40).[28] As Professor Gordon points out,[29] it is not surprising that the classical law 'shows no rooted objection to the passing of ownership without the transfer of possession' in view of 'the clear distinction made between ownership and possession'. The point at which an intention to transfer can be inferred is also the point at which the transferor ceases to hold as owner; in that he continues to have physical control, he does so on some basis other than that of owner.

The alternative form closest to an actual handing over is *traditio longa manu*, where an act of setting apart and making available the subject matter for the transferee to assume physical control is sufficiently well on the way to actual delivery to be recognised as such. The generic label *traditio ficta* ('fictional delivery') is also applied – possibly more appropriately – to the cases of the symbolical handing over of a key or some other token.[30] In classical law there was some difficulty in the notion of delivery wholly on the basis of the handing over of a symbol,[31] but this and the other seemingly diverse forms of *traditio ficta* may be rationalised on the basis of two elements: first, the presence of some act or event which may be identified as the act of delivery demonstrating the parties' intention, and second, a minimal physical requirement which may be no more than the identification of the subjects concerned for the purpose of delivery. Both elements are reflected in a well-known *Institutes* text (J.2.1.45): 'If someone sells goods deposited in a warehouse, the property in them passes to the buyer as soon as the keys of the warehouse are delivered to him.' This text envisages, at least, that the parties have identified, for the purpose of delivery, the merchandise in the warehouse and that they intend to transfer ownership in it.

Traditio brevi manu occurred where the transferee already held the thing but on a basis other than that of owner. The *Institutes* (J.2.1.44) speaks of the owner's intention being enough to pass ownership but, of course, where the transferee already holds the thing one may see the intention element acting on a pre-existing condition which satisfies the physical requirement.

Constitutum possessorium was the case in which the transferor continued to hold, but on a basis other than that of owner. It is doubtful that it could have been accepted merely on the basis of the transferor commencing to hold for and on behalf of the transferee. This would have been in open conflict with the rule that ownership does not pass by agreement alone (*traditionibus non nudis pactis dominia rerum transferuntur*).[32] The texts present considerable difficulty,[33] but what was probably required was that the transferor commenced to hold on the basis of a right consistent with the acquisition of ownership by the transferee – as in the case of the transferor retaining the thing under a contract of hire with the transferee. *Constitutum possessorium* and *traditio brevi manu* are related in that in both the critical intention requirement is satisfied by a change in the basis under which the thing is held. That in the former the transferor continues to hold while in the latter the transferee does is not a pertinent difference from the point of view of the physical element of delivery. In both, the subject matter, having been identified for the purposes of the change in the basis under which it is held, may be taken to be identified for the purpose of delivery. Of course, where the two modes do differ materially is that in the case of *constitutum possessorium* the thing is held by the non-owning party after the act of delivery, but this is a problem of policy relating to the interests of third parties rather than one of delivery.

The special requirement in respect of sale. Where delivery was pursuant to a contract of sale, ownership only passed on payment of the price or on the buyer giving security for payment unless the sale was on a credit basis (J.2.1.41). Justinian states that this rule is in accordance with natural law and that it originated in the *Twelve Tables*. The latter proposition is doubtful; there is no mention of the *Twelve Tables* in the statement of the rule in the *Digest* (Pomponius, D.18.1.19; Gaius, D.h.t.53). That the rule was Justinianic seems to be supported by Gaius' categorical statement (G.2.20) that title in a *res nec mancipi* passes on delivery whether the *causa* be sale, gift, or anything else.

The rule is consistent with the role of intention in derivative acquisition, because unless security has been given by the buyer or credit extended by the seller, the assumption must be that the seller, upon giving delivery, intends to pass ownership only on the basis of the price being forthcoming. Given that in the case of a cash sale the parties are subject to mutual obligations to deliver and to make payment, it is not inappropriate for the proprietary consequence of delivery to be restrained pending payment.

Provided the sale was for cash the unpaid seller could vindicate from the buyer or, indeed, from a subsequent party. Insofar as a subsequent party was innocent of the original buyer's failure to pay the price, he would be in a position to obtain a title by usucapion. However, having obtained the thing from a non-owner, he would not – pending acquisition – be protected against the owner (see section (e) below).

(e) Original acquisition: usucapion
(J.2.6; 4.6.4)

Usucapion was a process of original acquisition which facilitated the obtaining of a title through possession (see section (a) above). The device had a central role in the Roman property system. For a considerable period the formalities prescribed by the state law controlled the derivative acquisition of *res mancipi* (see section I(b) above). While this situation prevailed, a function of usucapion was to make a transferee owner where ownership would have passed had the necessary formalities been complied with. This was essentially a matter of rectifying a formal defect in title on the basis of a relatively short period of possession (J.2.6 pr; G.2.42) by a transferee who had acquired in a manner which satisfied the requirements of derivative acquisition (G.2.41). Another role of usucapion, fundamentally different from that of curing defective titles, was to make the transferee owner where he could not otherwise have acquired because his transferor was not in a position to convey ownership (G.2.43). Gaius notes that this obviated uncertainty without being unfair, in that the periods for usucapion were sufficient to protect a vigilant owner (G.2.44). As noted below, a point important to any policy justification was that stolen goods could not be acquired.

Gaius (G.2.42) gives the *Twelve Tables* as authority for the proposition that the acquisition by usucapion of moveables occurred in one year, while immoveable property was acquired in two years. In later law the periods were increased to three years for moveables and ten or twenty years for land, depending upon whether the owner was domiciled in the same or in a different province (J.2.6 pr). Land – especially unimproved land – was quite likely to be left unattended; an owner who did leave his land was seen to be entitled to greater protection where he was domiciled in another province.

As to the requirements of usucapion, an obvious one was that the thing had to be open to acquisition in the sense of being property which could be subject to individual ownership (G.2.48). In addition four substantive requirements – the active essentials of usucapion – had to be satisfied. First, there had to be a *iusta causa*; second, possession had to commence in good faith; third, the thing must not have been stolen or taken by violence; and fourth, there had to be uninterrupted possession for the requisite period.

The *iusta causa* requirement meant that possession had to follow from some act or event on the basis of which, other things being equal, the transferee would have acquired ownership. The relevant title in the *Institutes* (J.2.6 pr) opens with the statement that usucapion applies when a thing is bought, or received by gift or on the basis of any other just cause.

The need for a *iusta causa* is shown by considering the role of the requirement of good faith. This was satisfied by an honest, even if unfounded, belief that ownership had passed. The *iusta causa* requirement meant that the belief had to be on some basis which was sustainable as a matter of objective fact; accordingly, a mistaken belief would not suffice (J.2.6.11). Conversely, the existence of an appropriate *causa* was not enough if the transferee knew that in the circumstances he could not acquire. Provided that the transferee was in good faith at the time he assumed possession, the fact that he ascertained the truth before the completion of the period of usucapion did not prejudice his right to acquire.[34] Where, however, the *causa* was gift, possession had to be in good faith throughout the period, although Justinian abolished this special requirement (C.7.31.1.3 of AD 531).

The requirement that the thing must not have been stolen or taken by violence, which originated in the *Twelve Tables*, was an important limitation upon usucapion (J.2.6.2; G.2.45). The rationale was that the circumstances under which the true owner had lost possession warranted an exception to the policy of protecting a subsequent *bona fide* acquirer. The notion of a 'real defect' or *vitium reale* is brought out by a passage in the *Institutes* (J.2.6.3; see also G.2.45) which explains the thrust of the limitation. Noting that the thief or violent possessor would, in any event, be precluded on the basis of bad faith, the text continues: 'a third party has no right to usucapt, even after buying or acquiring them properly in good faith.' (J.2.6.3). The policy behind barring usucapion in the circumstances in question was taken to its logical conclusion in the recognition that the defect was purged when the owner got the property back or was in a position to do so but neglected to (J.2.6.8). The sources acknowledge that the effect of barring the usucapion of stolen goods was to produce a significant limitation upon the scope of usucapion of moveables;[35] but, as Gaius (G.2.50) notes, there remained 'other ways that someone can transfer a third party's property to another without the taint of theft and leave it possible for the recipient to usucapt the thing.' Gaius gives the example of an heir selling a thing lent to the deceased in the honest belief that it was part of the inheritance. This would not amount to theft and there would be no bar to acquisition by usucapion with the effect of terminating the right of the lender to the deceased. From a policy point of view there is a justifiable reluctance to allow usucapion where the owner suffered some act of deprivation. Where, however, the owner parted with the thing on a voluntary basis, even though not intending to convey ownership, the policy consideration is less compelling. Roman law appears to have recognised this.

What Gaius (G.2.53) describes as a liberal form of usucapion applied in the case of property forming part of an inheritance. This form of usucapion, known as *usucapio pro herede* (G.2.52), is said to have developed because the inheritance itself – as distinct from the items which comprised it – was subject to the shorter period of usucapion (G.2.54). Although in later law an inheritance could not be acquired as an entity, the shorter period continued to be applied to the items which made up a deceased's estate, including land (ibid.).

The circumstances of succession produced an exception to the rule of continuous possession. An heir could continue the period of possession commenced by a predecessor, despite knowing that he had no title to the property but, reasonably enough, only if the deceased's possession had commenced in good faith (J.2.6.12). Another exception to continuous possession was the recognition, in later law, of the possibility that a purchaser in good faith might add on the period of his predecessor in title provided the latter had also been in good faith (J.2.6.13). This concession could be seen as appropriate in the context of the extended periods of acquisitive prescription which applied in later law.

In principle a usucaping possessor was protected to no greater extent

tacking

than any other possessor. The status quo of possession was protected against an unlawful act of dispossession but, of course, the possessory interdicts (see section II(c) above) were only available against an immediate dispossessor. Prior to the completion of usucapion there was, in principle, no defence against a *vindicatio* brought by one who could prove a right of ownership; *a fortiori*, there was no basis for recovery if the thing came into the possession of another (J.4.6.4; G.4.36). The recognition of a limited proprietary title pending the completion of the period of usucapion came in the late Republic. The Publician action – a real action based upon the fiction that the period of usucapion was complete – gave the crucial general right to recover (J.4.6.4; see also this section below, and Chapter 6, section V below).

A party usucaping only because of a formal defect in the process of acquisition – typically the informal transfer of a *res mancipi* (see section (b) above) – was protected as if he were owner. That such a 'bonitary owner' came to be protected against the titular owner is hardly exceptionable given that all that was absent in the process of derivative acquisition was compliance with formality. Accordingly, a bonitary owner had a defence that the thing had been sold and delivered (*exceptio rei venditae et traditae*) against the vindicating owner (Hermogenian, D.21.3.3). Probably more importantly, the bonitary owner could recover the thing from any party as if he were the owner.

A *bona fide* possessor was one who honestly believed in his right to the thing but whose title was deficient in substance in that he had acquired from a non-owner. Under the Publician action a *bona fide* possessor could recover as if the period of usucapion was complete but, of course, not from the true owner, who could defend a Publician action *in rem* by resort to the defence of ownership (*exceptio iusti dominii*: Paul, in Papinian, D.6.2.16; Neratius, D.h.t.17).

(f) Original acquisition: occupation
(J.2.1.12 - 17, 39, 46 - 48)

In various situations title could be obtained by taking possession of a thing which was open to acquisition but presently unowned (G.2.66). From the point of view of the principles of private law the most important instances of occupation are those relating to wild creatures and abandoned things.[36]

The essentials of occupation are as follows. First, the thing must be open to acquisition in being susceptible to private ownership and unowned. Second, there must be a reduction to possession with the intention to acquire ownership by the occupier. Because the intention to acquire can usually be inferred from the act of taking, the issue tends to be whether there has been an effective taking (J.2.1.12). This, in turn, resolves itself upon whether the thing can be said to have come into the power of the claimant (Proculus, D.41.1.55).

With regard to naturally wild creatures[37] the principle was well established that the first taker acquired, and this was so regardless of where the reduction to possession occurred. The point was noted (J.2.1.12) that while a landowner had the right to exclude a poacher, the latter nonetheless acquired any game he took. Even where a permanent habitat was established naturally, wild creatures did not fall within the proprietary domain of the landowner concerned. On this basis a swarm of wild bees belonged to no one (*res nullius*) and was open to acquisition wherever it hived (J.2.1.14). Logically, its honey was *res nullius* and not susceptible to theft (ibid.).[38]

There was some debate as to whether a seriously wounded animal could be deemed to be taken or provisionally taken while the hunter remained in pursuit. Justinian (J.2.1.13) rejects this extended notion of a taking on the basis that too much uncertainty prevailed pending a final reduction to possession. The ownership of a wild creature was lost when it regained its natural liberty and, of course, it then became open to acquisition again (J.2.1.12; G.2.67). The captor's right was deemed terminated when the animal was no longer in sight or, though within sight, when pursuit was difficult (J.2.1.12). However, certain wild creatures, once acquired, remained owned, provided that, even though they did not remain permanently on the owner's property, they retained the habit of returning or 'homing instinct' (*animus revertendi*). Pigeons and bees are the best-known examples, but the sources also mention peafowl and deer (J.2.1.15; G.2.68). Upon the loss of its homing instinct a creature of this class, having left the owner's domain, became again open to acquisition on the basis of being *res nullius* (J.2.1.15).

As the ultimate complex of rights and powers over a thing, ownership necessarily encompasses an unlimited right of disposal, including self-imposed termination by act of abandonment. The *abusus* element in the *usus*, *fructus*, *abusus*[39] breakdown of constituents includes this right. Upon being abandoned the now unowned thing (a *res derelicta*) becomes open to acquisition by occupation. In speaking of the abandoner immediately ceasing to be owner, Justinian (J.2.1.47) does not subscribe to the Proculian view that the ownership of an abandoned thing was not lost until another party assumed possession (Paul, D.41.7.2.1). Confusingly, however, the opening words of the relevant text (J.2.1.47) indicate analogy to the subject of the preceding paragraph, the transfer on a derivative basis in the giving of largesse to indeterminate persons (J.2.1.46).

The fact that a thing is physically left does not, of course, necessarily mean that the owner intends to terminate his right. Ownership is not lost in items put overboard to lighten a ship in peril (J.2.1.48; see Chapter 5, section V(b) below). Justinian says that this situation is analogous to that of '[things] which drop out of moving vehicles without their owners' knowledge' (ibid.). The latter case does not involve an act by the owner, but the relevance of the analogy lies in the fact that neither case is consistent with an inference that ownership has been abandoned.

Treasure trove is usually dealt with as a separate mode of acquisition, but neither its practical importance nor its significance from the point of view of legal principle justify this. It may be noted here that hidden treasure is not a *res derelicta*: one can hardly infer an intention to abandon ownership. Allocating rights in a find of apparently unowned valuables is a matter of policy for any legal system. The emperor Hadrian (AD 117-38) laid down rules which are set out in the *Institutes* (J.2.1.39). Where a man found treasure on his own land he could keep it; if the find was made accidentally on the land of another the finder and the landowner shared the treasure; if the find resulted from a search the landowner acquired the entire treasure. It may be noted that the *fiscus* only benefited if the state was the landowner (ibid.).

(g) Original acquisition: accession
(J.2.1.20 - 24, 26, 29 - 34)

The Roman law of property recognised, in various situations, the notion of an accessory item being or becoming part of a principal thing to which it could be said to belong (*accessio*). The most important application was the case of the union of two separate items, giving a single thing on the basis of the accessory being subsumed under the principal. Where this occurred the owner of the principal item acquired the accessory when the union came into being. Another aspect of the concept was the recognition that, in principle, the produce or progeny of property, even though a separate thing, belonged to the owner of the principal or parent thing. The important difference between these two applications of accession was that the first type reflected a proprietary *fait accompli* whereas, in the second case, because the principal and accessory things were separate, there could be an allocation of rights on a basis other than that the accessory follows the principal.

Various cases concerned with natural changes to land (whether through imperceptible increment, a river altering its course or the force of flood water) are dealt with in the sources, but these are somewhat specialised applications of the concept (J.2.1.20 - 24; G.2.70 - 72). The most important application to land was concerned with the erection and improvement of buildings (*inaedificatio*). What is sometimes referred to as the problem of 'fixtures' remains, in modern law, the aspect of accession most likely to produce controversy.[40]

The case of accession to land does not involve any primary difficulty in the identification of the principal element in the union of two things. Regardless of the value factor, the land is principal: 'anything built on land becomes part of the land' (J.2.1.29; see also G.2.73). Roman law gave effect to the principle of accession in the case of *inaedificatio* by a somewhat unsatisfactory compromise. The consequence of accession was taken to be that the acceded item could not be claimed as long as the structure stood, but the 'owner' was entitled to it upon the building being dismantled. This equitable dormant interest – which Lee[41] rightly labels 'barren consolation' – is a distortion of principle. No right can be maintained following accession because the very effect of accession is to terminate the separate existence of the acceding thing.[42]

The action for beams set in (*actio de tigno injunto*) provided for the possibility of compensation in the case of the use of stolen materials in the erection of a building (J.2.1.29). It would appear that this action was proprietary in that it prevailed against one who used the materials but was innocent of the fact that they had been stolen (Ulpian, D.47.3.1.2). In view of the proprietary character one would assume that the action should have been available against the innocent purchaser of a building in which stolen materials happened to have been incorporated.[43]

Accession to land also occurred in the case of anything which took root, regardless of the circumstances in which the planting or sowing had taken place (J.2.1.31). Gaius (G.2.74) regards this as a stronger form of accession than *inaedificatio*. This does not mean that there are degrees of accession but simply that it is easier to identify the phenomenon in certain forms; clearly, the rooting criterion simplifies matters.

Where the builder or planter was a *bona fide* possessor of the land he was in a relatively strong position from the point of view of compensation. The defence of fraud (*exceptio doli mali*) was available as a defence against a party seeking to vindicate the property concerned. It allowed a *bona fide* possessor to retain the principal thing pending compensation by the claimant to meet the expenditure incurred in the accession (J.2.1.30; G.2.76). This right of retention was proprietary in that a real interest inured to one who improved property possessed in an honest belief that it was his.

Special rules were developed in respect of writing and painting. The corporeal entity of any sort of written character acceded to the paper or other material upon which it was inscribed (J.2.1.33; G.2.77). This result followed even in the case of gold lettering (as Nicholas[44] notes, an instance of the accession of a more valuable item to a less valuable one). Of course, in respect of any form of writing, the question of ownership of the incorporeal composition is likely to be of greater importance than the issue of who owns the actual corporeal characters. The notion of 'copyright' does not seem to have been grasped in the *Institutes* (J.2.1.33), which gives ownership of a poem, a history, or an oration to the owner of the paper on which the item is written. The obvious qualification is that the owner of the paper owns the particular corporeal entity based on his paper but not the incorporeal composition, which remains the property of the author.

The case of the painting of a picture produced some controversy, but the better view was taken to be that the canvas or board acceded to the picture

(J.2.1.34; G.2.78). In the case of a picture (and in contrast to a literary or musical composition) the incorporeal aesthetic quality is inseparable from the corporeal entity. On this basis there would appear to be a case for the distinction between writing and painting which Gaius (G.2.78) considered to be something without a satisfactory foundation.

As with other instances of accession, a *bona fide* possessor who had put money into the thing was protected by the *exceptio doli* against the vindicating owner (J.2.1.33, 34; G.2.78). One can hardly see this as having more than a theoretical role in respect of characters on paper. In the case of a picture, in the unlikely situation of the painter having lost possession, he could only vindicate after first compensating the former owner of the canvas if, equally improbably, the latter was in good faith (J.2.1.34).

In all cases of accession, a deprived party might have a personal claim against one who had procured the accession by wrongful act, or was otherwise in bad faith (J.2.1.26; G.2.79). However, a claim was not admitted merely on the basis of the unjustified enrichment of the owner of the principal thing.[45] These issues, of course, are matters of the law of obligations rather than property (see Chapter 5, section IX(b) below).

(h) Original acquisition: specification
(J.2.1.25)

The production of a distinct new entity through processing or manufacture involving the use of the property of different parties (*specificatio*) necessitates a decision as to the allocation of rights in the new product. Likewise, a decision is needed in the case of a new thing made by one who did not own the raw materials, as where 'one man makes wine, oil, or grain from another's grapes, olives, or corn' (J.2.1.25). But the adoption of a solution was somewhat controversial; as Gaius (G.2.79) points out, one school of jurists preferred to give effect to the interest of the owner of the materials, while the other sought to allocate the end product to the maker on the basis of his contribution. Justinian (J.2.1.25) resolved the matter by a compromise. The owner of the materials was preferred if the final new product could be reduced to its constituents; if it could not be so reduced, the maker was preferred. Hence, *a fortiori*, where the maker of an irreducible thing had also contributed materials he was taken to own the end product.[46]

Where specification applied, ownership of the new entity was allocated to the maker to the exclusion of any proprietary interest vesting in the former owner of a constituent or ingredient element, the separate identity of which terminated in the act of making. In the improbable event of a party so deprived coming into possession in good faith, he would be in a position to resort to the *exceptio doli* and resist the maker's *vindicatio* pending payment of compensation.

The better view is that the proprietary consequence of specification followed even if the maker acted in bad faith; in principle, the act of specification is simply an instance of the consumption of the property of another. But, of course, personal remedies – similar to those applying to an act of accession in bad faith (J.2.1.26; G.2.79) – were potentially applicable where the maker was in bad faith.

(i) Original acquisition: mixture and fusion (J.2.1.26 - 28)

The mere fact of the mixing together of solid items (*commixtio*) did not lead to a change of ownership. No change occurred where sheep owned by different parties came together in a single flock (J.2.1.28). A flock is not a single entity to the exclusion of the independent existence, as separate items of property, of the individual sheep.

Where the nature of the items mixed makes it impossible to identify the respective contributions – as in the case of the mixing of grain – the property interests of the owners concerned would be translated into rights to rateable shares (J.2.1.28). As Justinian notes, at the end of the day, the determination of the respective shares might have to be a matter of judicial discretion.

Where the act of bringing together produced a single entity that could not be reduced to constituents, a redefining of property rights was necessary. This case of fusion (*confusio*) is typically represented by the mixing of liquids or the fusing of metals into an irreducible substance (J.2.1.27). Where the respective owners consented to the mixing, their agreement would determine the basis under which the corporate entity was owned; if they had not agreed on the proportions, the result would be that ownership was common (ibid.). The better view, although not reflected in the *Institutes*, is that the merging of substances by agreement is a matter of derivative rather than original acquisition with the actual act of mixing amounting to delivery.

Where the concepts of accession or specification apply, they do so to the exclusion of rules applicable to mixing and fusing. Accession and specification are instances of acquisition on particular rational bases; the allocation of rights in a merged entity is a compromise which need not be embarked upon if there is a more appropriate basis under which ownership can be allocated.

(j) Original acquisition: fruits (J.2.1.35 - 38)

Roman ownership, comprising the classes of rights designated *usus*, *fructus*, and *abusus*, was open to being split up according to these classes. The breakdown meant that the law could allocate the rights to the use and to the fruits (*usus* and *fructus*), while the owner remained owner by retaining the right of disposal (*abusus*) which, of course, was fundamental to ownership (see section II(a) above).

The fruits of property included the produce of land and the offspring of animals (J.2.1.37) and, by logical extension, any financial return on property (Ulpian, D.22.1.36). These two classes of fruits are labelled, respectively, natural and civil fruits. The usual and normal position was that an owner had the right to the fruits of his property, and the fact of their coming to have a separate existence by separation from the parent entity was an irrelevant consideration. But Roman law recognised that upon separation it was possible for a party other than the owner to acquire the fruits.

The most obvious case of acquisition of fruits by a non-owner was that of the usufructuary (J.2.1.36; see section IV(b) below). One who was not owner but had the right to the use and fruits of a thing became owner of the fruits by the act of gathering or harvesting (*fructuum perceptio*). Thus the holder of the real right of usufruct of a parcel of land acquired the fruits of the soil by an act of taking for himself. The requirement of an act of taking meant that upon termination of the usufruct any fruits not taken passed with the parent property (ibid.).

The right to fruits which was part of usufruct applied in other instances of the passing of lesser real rights. Certain forms of lease and secured tenure fell into this category (J.2.1.36).[47] However, in the case of tenancy on a purely contractual basis the tenant did not acquire the actual rights of *usus* and *fructus* but only a right against the owner to have the use and the fruits of the property. As Nicholas[48] shows, the tenant, in this case, actually acquired on a derivative basis, by a form of constructive delivery, in that acquisition of fruits was on the basis of the owner's consent (Africanus, D.47.2.62(61).8).

Fruits were acquired by a *bona fide* possessor of the parent thing upon separation from it (*fructuum separatio*; see Paul, D.41.1.48 pr; J.2.1.35). The *bona fide* possessor was one who honestly believed that he had a right to the fruit-bearing property but, of course, a claim to possess in good faith would only be credible if it was supported by the circumstances. This meant that, in practice, a *bona fide* possessor was one who had obtained the thing in circumstances consistent with the acquisition of ownership, as is apparent from the way the *Institutes* poses the question: 'Suppose you believe that someone is owner of a piece of land when in fact he is not, and you buy it from him in good faith or, still on the assumption of good faith, receive it as a gift or on some other legally sufficient ground' (J.2.1.35).

Justinian (ibid.) identifies the policy basis of the acquisition of fruits by a *bona fide* possessor as a reward for the trouble taken in cultivating the land. Certainly, the rule does appear to be policy-based because, as a matter of property, the fact of separation does not of itself signal a change of ownership without reference to the owner of the parent entity.[49] Roman law was consistent in applying policy considerations to the possessor's right to fruits. Thus, a *mala fide* possessor had no right to, and indeed was accountable for, consumed fruits. Moreover, the policy basis behind allocating fruits to a *bona fide* possessor was subject to the logical limit that the true owner, on proving his right to the parent thing, should be allowed to recover any unconsumed fruits (ibid.).

IV. Lesser Real Rights

(a) Introduction

Roman law identified certain rights which were equal to ownership in also having the quality of being universally applicable. These rights, like ownership, were protected by remedies available against the world at large. The rights in question were 'lesser' in the sense that their scope was less than that of ownership. That the rights had the same standing but included less than ownership follows from the fact that they were actually rights, or packages of rights, removed from the full complement of rights comprising ownership.

The various forms of lesser real rights had the common feature of placing a limitation upon the right of ownership from which they were abstracted. Thus they came to be known as real rights over the property of another (*iura in re aliena*).[50] There were two basic types of limitation. One, already referred to, in the most common form of usufruct (see section III(j) above) took off the *usus* and the *fructus* aspects, leaving ownership constituted by a right of disposal (*abusus*) of the bare title. These forms amount to a limitation in the sense of the complete or partial removal of the beneficial aspects of ownership. The other type of limitation left ownership essentially intact – the core rights to use, fruits, and disposal were retained – but curtailed some particular aspect of the owner's rights: for example, the case of land subject to a servitude of right of way. The various forms of real security gave to another a real interest protected by limiting the owner's right of disposal in some way or other.

The main forms of lesser rights will be dealt with under the headings of personal servitudes, praedial servitudes, and real security. These lesser rights fit the dogmatics of the concept of ownership in Roman law. Two forms of holding immoveable property, *emphyteusis* and *superficies*, gave a right approximating ownership, but were in a sense alien to the Roman system. Although they were not wholly compatible with the structure of Roman property, the rights concerned did not deny an ultimate right of ownership and, in this sense, they may be seen as lesser rights. Therefore *emphyteusis* and *superficies* will be dealt with in section IV(e) as 'other lesser rights'.

(b) Personal servitudes
(J.2.1.38; 2.4 - 5)

Usufruct. This was the most important form of personal servitude and the other forms are derivatives of it. The personal servitudes were so designated because they featured a vesting of all or part of the beneficial interest of the right of ownership in a person other than the owner, the right being personal to that person and so inalienable.

Justinian's definition (J.2.4 pr) identifies the nature of usufruct as the removal to someone other than the owner of the rights to the use and the fruits, subject to the substance of the property remaining unimpaired. The obvious necessity that the thing be preserved[51] disallowed action by the usufructuary which might affect the substance of the property. The owner retained the bare property (G.2.30) and the right of disposal (*abusus*) of it. But the property had to be preserved, and the owner accordingly had no more than a right of alienation of the bare property; in practice, however, the usufructuary had possession and was in a position to protect the property.

The requirement that the usufructuary had to maintain the substance of the property, so that its capacity for use and for the production of fruits was undiminished, was given effect according to the nature of the property concerned. As Justinian notes (J.2.1.38), the usufructuary of a flock had to maintain the numbers, and, in the common case of a usufruct of land, dead vines or trees had to be replaced. If the usufruct was over a timber estate, mature trees could be cut as fruits but there would have to be periodic plantings to maintain the substance of the property.[52] In the case of a quarry or mine, the rule against reducing the substance precluded the annual removal of more than a reasonable quantity of the stone or mineral concerned, it being a basic principle that the usufructuary had to exercise his right in a reasonable manner.[53]

The prerequisite that the substance remain unimpaired had implications in respect of consumables. Thus, while in principle there could be no usufruct of a quantity of a substance consumed by its use, it came to be competent to allow a usufruct provided security was put up for the restoration of the amount concerned; this also applied to a usufruct of money (J.2.4.2).

A usufruct over a *res mancipi*, constituted *inter vivos*, was created by *in iure cessio* or as a reservation in the context of *mancipatio* with the transferor retaining the usufruct while bare title went to the transferee (see section III(c) above). In later law constitution *inter vivos* was by pacts and stipulations (J.2.4.1). In practice, however, most usufructs came into being on the basis of a testamentary act. Usufruct made it possible for a testator to provide for a surviving party during his or her lifetime with the property retained intact to pass in full to the bare title holder – the ultimate beneficiary – on the death of the survivor. Where a legacy in a will left the usufruct of property to a particular party, bare ownership would automatically go to the heir (ibid.).

A usufruct was terminated by the death of the usufructuary or by deprivation of civil law right-bearing status (J.2.4.3). The usual form of usufruct was for the lifetime of the usufructuary, but a usufruct for a fixed term ended upon the death of the usufructuary before the expiry of the period. Upon termination, the rights of *usus* and *fructus* automatically reverted to the owner (J.2.4.4). Usufruct ended if the rights of *usus* and *fructus* were ceded by the usufructuary to the owner and, equally, if the owner ceded to the holder of the usufruct the residual right of ownership (J.2.4.3). There could be no usufruct in a situation in which all the relevant rights resided in the same party. Given the nature of usufruct it was only logical that if the thing concerned was destroyed or rendered useless the usufruct terminated. A usufruct of a house ended if the house was burnt down or destroyed by earthquake, or even if it fell into a state of ruin (ibid.).

As Nicholas[54] shows, there is only a superficial similarity between the usufruct in Roman law and the separation of beneficial interest and title on the basis of equitable and legal ownership as developed in English law. One might go further and note the difference between usufruct and the trust device derived from English law and used in various systems to separate beneficial interest and title. The difference follows from the fundamentally distinct starting points applying to the two forms. The Roman usufruct is a natural by-product of a unitary right of ownership made up of separable components (the rights to use, to fruits, and of disposal). The English trust, on the other hand, developed on the basis of the distinct forms of legal and equitable ownership. It may be noted that a Roman form of trust is the *fideicommissum*, but this was a device of the law of succession which made possible the settlement of property within the prevailing notion of ownership.[55]

Usus. This was the right to the use of a thing without access to its fruits. It was brought into being and terminated in the same way as the more extensive right of *usufruct* (J.2.5 pr). *Usus* was not appropriate to the case of unimproved land, but in theory the usuary had a right to a part of the produce of the land for the limited purpose of his daily use (J.2.5.1).

Habitatio. This was a variation of *usus* extending the right to the occupation of a house to allow the beneficiary to let it (J.2.5.5).

(c) Praedial servitudes
(J.2.3)

Praedial servitudes regulated the relationship between separate parcels of land with regard to matters of mutual relevance having a bearing upon the use or enjoyment of one of the properties. That praedial servitudes were only to do with immoveable property is made clear in the *Institutes*

(J.2.3.3), where it is stated that only a landowner could acquire the benefit of or be burdened with such a right of servitude. The most important areas of application were in relation to access, water (including drainage), support, and light.

A praedial servitude operated so as to place a particular limitation upon some aspect of a landowner's right (see section (a) above) for the benefit of the owner of another, usually adjacent, property. Although the benefit/burden inherent in a praedial servitude affected the use of the properties concerned by the parties involved, the right/duty relationship was not one between individuals but, rather, one between parcels of land. Because praedial servitudes could only be concerned with the relationship between two parcels of land, the right/duty reciprocity had to relate to the actual use of the land rather than merely to some peripheral activity made contingent upon the existence of rights to land. The right/duty relationship was associated with the land to the extent that the law speaks in terms of the dominant and the servient properties (*praedium dominans* and *praedium serviens*).[56]

Properly constituted a praedial servitude gave a real right over a particular parcel of land applying in favour of another parcel.[57] Because it was a right in favour of one parcel to which another parcel was subject, and regulated some aspect of the relationship between the properties concerned, a praedial servitude endured regardless of a change of ownership in either property. The efficacy of a contractual right between owners would be limited to the period of their coincident ownership. A real right in favour of the dominant owner would have to be reconstituted in the event of a change of ownership in the dominant property.

Types. The primary division was between rustic and urban servitudes and the distinction was relevant to creation. Gaius (G.2.14) speaks of 'rights attached to urban and rural lands', but the *Institutes* (J.2.3.1) states that urban servitudes apply to buildings even if the land is in the country.

The so-called rustic servitudes were primarily to do with access, with the most important providing for a right of way over the servient land. The three forms commonly mentioned are passage, drive, and way (*iter*, *actus*, and *via*; J.2.3 pr). The first gave only a right of pedestrian access, the second extended this to a right to drive livestock, while the third allowed access by vehicles also and included standard specifications concerning the width and surface of the road (Gaius, D.8.3.8). The servitude of aqueduct (*aquae ductus*; J.2.3 pr)[58] gave a right to lead water over the servient land and, of course, this was a form of access. Other rustic servitudes mentioned in the *Institutes* (J.2.3.2) are the rights to draw water, to water cattle, to pasture, to dig sand and to burn lime. Buckland[59] regards these forms as a species distinct from the rights of access and draws an analogy with the difference between 'commons' (or 'profits') and 'easements' of English law but, of course, the Roman system did not know this distinction.[60] The category of rustic servitudes represented the needs of Roman rural land-ownership rather than any dogmatic integrity but, this said, they do all reflect the general notion of reciprocal benefit/burden applying between two parcels of land.

The so-called urban servitudes were essentially to do with the necessary mutual relationship between adjacent properties developed by building. Thus the servitude *oneris ferendi* provided for the duty to support an adjoining building. Although, in principle, a servitude could not impose a positive duty (Pomponius, D.8.1.15.1), it seems hardly 'remarkable'[61] that the servitude of support necessarily involved the owner of the servient property in the positive duty of maintaining his supporting structure. Moreover, there does not seem to be any difficulty in this obligation passing, as an essential part of the servitude, to a new owner of the servient tenement – the consequence, of course, of the right being a real one.[62] A servitude also concerned with support allowed a beam from a building on the dominant property to be inserted into a building on the servient property (J.2.3.1). Various urban servitudes were concerned with drainage and the flow of water. A servitude providing for drainage from the dominant property through the servient one was one of the common forms.[63] The *Institutes* (J.2.3.1) mentions the servitude to receive a flow of water diverted by a spout and this would appear to be concerned with drainage. Like the servitudes relevant to support, the duty to receive dripping rainwater from a neighbour's eaves is an indication of congested urban development. In principle a landowner could build to any height but there could be a servitude which limited building on the servient land in the interest of the light reaching the dominant property (ibid.).

Creation and termination. In Gaius' time, insofar as the land concerned was *res mancipi*, rustic servitudes were created by *mancipatio* or *in iure cessio* while urban servitudes could only be created by the latter means (G.2.29). Given the potential importance of servitudes in the urban context it may have been thought appropriate that creation should be by official act rather than through the private mancipatory process (see end of section III(c) above). In any event, in later law praedial servitudes were created by the informal process of pacts and stipulations, which was the prevailing system in the provinces (J.2.3.4; G.2.30). Testamentary provision for a praedial servitude was competent (J.2.3.4) but far less likely than in the case of a personal servitude (see section (b) above). Through a somewhat strained development[64] Roman law reached the logically apposite position that praedial servitudes could be acquired by prescription (C.7.33.12.4 by Justinian; C.3.34.1 of AD 212). Where a right of servitude had been continuously exercised neither by the use of force, nor clandestinely, nor on the basis of a revocable grant (*nec vi, nec clam, nec precario*), the right was acquired after ten or twenty years depending upon whether the servient owner was domiciled in the same or in another province.[65]

Where the dominant and servient properties came into common ownership, the essential basis of a praedial servitude was lost and the servitude was extinguished. If the properties concerned came again to be separately owned, the servitude did not automatically revive but had to be reconstituted (Ulpian, D.8.4.10). A praedial servitude came into being to serve some necessary purpose between properties and would normally be perpetual. Exceptionally, a praedial servitude might be constituted for a limited period because either the dominant or the servient tenements were subject to a limited right. Renunciation, equally, would be likely only where changed circumstances made the servitude unnecessary. In classical law renunciation, like creation, was effected by *in iure cessio*, but in later law informal agreement was sufficient.[66] Through negative prescription a praedial servitude was lost by non-user for the same ten- or twenty-year periods as applied to acquisitive prescription (C.3.34.13 of AD 531). In the case of an urban servitude there had to be an obstruction which prevented the exercise of the servitude for the period.

(d) Real security
(J.3.14.4; 4.6.7)

Real security is the generic term for real rights created or recognised to secure the performance of an obligation (see Chapter 5, section II(e) below). A creditor in a purely contractual relationship with a debtor is vulnerable to the debtor's default. The object of real security is to strengthen the creditor's position by giving him a property right in something owned by the debtor, whether the item in respect of which the debtor is obliged to the creditor or some other thing. An effective real security should mean that the creditor can recover what is due to him regardless of the fact that the debtor's liabilities exceed his assets. This puts him at a considerable advantage over an unsecured creditor who will necessarily receive less than his due if the debtor cannot meet his total liabilities.

A feature of an economy of any sophistication is the need by various sectors to be able to raise funds. The question of credit may arise in an individual consumer transaction or in a programme of industrial or commercial development. Credit necessarily raises the issue of security. A legal system must provide appropriate forms which allow the optimum use of different forms of property as security. There must be a balance between protection of the creditor and fairness to the debtor.

These general observations concerning the role of security are broadly applicable to Roman law.[67] Over the development of the system, real security occurred in three basic forms: first, by a transfer of ownership from debtor to creditor; second, by a transfer of possession from debtor to creditor; and third, by a charge on the property which invested the creditor with a right precluding free disposal by the debtor. The three forms may be labelled *mancipatio cum fiducia*, *pignus*, and *hypotheca*.[68]

Mancipatio cum fiducia. This involved a conveyance of the property by *mancipatio* or *in iure cessio* so that title passed to the creditor. Concomitant with the conveyance was an agreement which would normally provide for reconveyance to the debtor upon his satisfaction of the debt. This obligation aspect of the parties' relationship was brought into being by an undertaking known as *fiducia* which operated as an appendage to the conveyance.[69] The critical feature of the composite device of *mancipatio cum fiducia* was that the creditor became absolute owner of the property while the debtor was left with a contractual claim against the creditor for reconveyance. The significant implication of this was that the creditor could give an unimpeachable title while the debtor had only a personal claim against the creditor. The word *fiducia*, applied to the parties' agreement as to reconveyance, implied a relationship of trust. However, this meant no more than that the breach of a *mancipatio cum fiducia* agreement attracted severe penalties (G.4.182). The law recognised the vulnerability of the debtor at the instance of the creditor produced by this form of security. However, for the security to be meaningful the creditor had to get title – including the right of disposal (*abusus*) – and the only alleviation to the debtor's vulnerability was by way of disincentive to any abuse by the creditor. The modern mortgage does not give title to the creditor but merely makes the debtor's right of disposal subject to the priority of the creditor's right. This is achieved by preventing the transfer of ownership by the debtor before cancellation of the mortgage by the creditor.

Manicipatio cum fiducia had a long history in Roman law and only disappeared when the formal modes of transfer went into obsolescence.[70] During its centuries of use there was no change in the basic feature of the creditor obtaining a right of ownership. But developments ameliorated the debtor's position by controlling the relationship between debtor and creditor. However, the most stringent controls of the creditor's right of sale could not prevent a transfer of ownership to a third party by a creditor acting in breach of his obligation to the debtor.

The debtor could recover ownership by a special form of usucapion, known as *usureceptio*, which required only one year's possession, even in respect of land (G.2.59). Where the debt had been paid, usucapion would proceed automatically on the restoration of possession (G.2.60): hardly a concession, given that 'as a matter of equity [the debtor was] on a par with a bonitary owner'.[71] Significantly, there could be no usucapion, pending payment of the debt, where possession had been restored on a 'lease-back' or precarious basis (G.2.60). This, of course, was in keeping with the notion of usucapion as the obtaining of title through possession on a basis consistent with the acquisition of ownership (J.2.6 pr; see section III(e) above). The role of usucapion should not be overestimated. That the debtor was in a relatively strong position if he had paid and had possession was irrelevant from the point of view of possible abuses by the creditor in possession.

Pignus. The simple device of handing over the possession of a thing to a creditor came to be a form of security with the development of the protection of possession as an independent right. Although the creditor did not 'possess as owner' (see section II(b) above), possession was – at least in the developed law – on the basis of the exercise of a right to the thing in the event of the debtor's default and, to this extent, it was distinguishable from possession on the basis of a contract – for example, hire or loan – which unreservedly acknowledged the right of another.

Justinian (J.3.14.4; see also Chapter 5, section II(e) below) deals with *pignus* in his treatment of contractual obligations. The rights and duties pertaining between creditor and debtor are matters of obligation, but the purpose and effect of the handing over of property as a pledge is proprietary. Although the debtor does not part with the right of disposal, the creditor receives property insofar as the right of possession is an integral part of ownership. Moreover, the creditor's right was protected by the normal possessory interdicts (see section II(c) above) as well as by a special interdict (J.4.6.7), remedies which were real in the sense of being directed to the thing and in principle available against anyone who held it. From this it would seem to be arguable that *pignus* was a *ius in re aliena*, but the Romans, apparently, did not regard it as such.[72]

A factor which limited the utility of *pignus* was that it was not competent to give a right of possession without handing the thing over to the creditor. This tended to limit its application to things which did not have a high utility value. In what would appear to be a response to this problem, a form known as *antichresis* provided for the deduction of fruits from interest (Marcian, D.20.1.11.1). *Mancipatio cum fiducia* had the potential to avoid the problem of the debtor losing the use of the thing concerned. Because the security was based on the creditor's title, the thing could be returned to the debtor to hold on a contractual or precarious basis (G.2.60). There is authority[73] that this was also competent in the case of *pignus* in respect of land, but the point is that a creditor would be less likely to be prepared to run the risk of restoring the thing to a debtor who had not given up title.

Although *pignus* came to envisage a possible sale in liquidation, it did not itself give a right of sale nor would one expect this from a mere transfer of possession. In earlier law its primary purpose may have been to induce payment rather than provide for the problem of non-payment.[74] However, a right of sale could be agreed and one came to be implied. Gaius (G.2.62), in dealing with the exceptional cases in which a non-owner could pass property, noted that a creditor could alienate pledged property even though it was not his (G.2.64). Gaius suggests that the explanation for this was 'that he alienates the pledge with the debtor's authority, previously given by a term of the contract allowing the creditor the right to sell on failure to repay'.

Hypotheca. In the *Institutes* (J.4.6.7) pledge (*pignus*) is distinguished from mortgage (*hypotheca*): the former applies to the case of a thing handed over to a creditor, the latter to the case of security in a thing created by simple agreement without delivery.

The development of a form of real security relying neither upon a transfer of title nor of possession occurred through the general application of provisions that had been particular to the relationship of landlord and tenant. The essential catalyst of this development was the problem – endemic to *pignus* – of the requirement of a handing over of actual possession to give an effective security. Plainly enough, this was wholly inappropriate in the context of any enterprise in which the means of production was the assets potentially applicable as security. Handing over animals and agricultural implements was hardly an appropriate way of giving security for the payment of rent by a tenant farmer. What was needed was some form of non-possessory security.

To remedy the problem, the praetor gave the landowner an interdict to enable him (upon the rent becoming due) to get possession of items pledged by a tenant as security (J.4.15.3). This was a remedy only available against the tenant; the subsequent granting by the praetor of an action (J.4.6.7) on the basis of which the thing could be recovered from a third party was the development which gave a real right. But as Nicholas[75] notes, the recognition of a *ius in re aliena*, purely on the basis of agreement and without any form of delivery, 'ignored the line between contract and conveyance'. Given that the landowner was invested with a legal right, he could leave the tenant in possession of the items essential to agricultural production – or of the furniture in the case of the lease of a house.

In time the development which had occurred in respect of landlord and tenant came to apply to a non-possessory security arrangement in any context. In the result, *pignus* and *hypotheca* were assimilated, so much so that according to one *Digest* passage (Marcian, D.20.1.5.1) it was a case of different words being applied to the same concept.

The development of non-possessory security made possible the creation of successive security rights in property remaining in the possession of the debtor. In a bankruptcy situation, after any privileged claims (typically that of the *fiscus*) had been met, ranking was on the basis of chronological priority (C.8.17(18).2 of AD 213). In later law problems began to arise through the incidence of multiple non-possessory security rights in a particular parcel of land. To combat this the emperor Leo (AD 457-74) introduced a preference for security agreements registered in court or subscribed to by three reputable witnesses (C.8.17(18).11.1 of AD 469). But without a system of recording, the holder of a subsequent privileged security interest would not necessarily be aware of an existing privileged interest which, of course, would take priority. Leo's reforms encouraged creditors to subscribe to the prescribed formalities. Between privileged creditors there remained a chronological priority, although it probably was a clearer one based on recorded dates.

A further factor concerning security was the incidence of tacit or legal *hypothecae*. These were preferences in favour of particular creditors, not created by agreement but simply recognised by the law in various circumstances in which, as a matter of policy, it was seen as appropriate for the claim to be secured. Tacit *hypothecae* were either special, against particular property, or general, against all of the debtor's property. An example of the former was the landowner's *hypotheca* over a tenant farmer's standing crops as security for rent (Pomponius, D.20.2.7 pr). An example of the latter was the *hypotheca* over all of a guardian's own property to secure a ward's claim in respect of the tutor's administration (C.5.37.20 of AD 314). A creditor who believed that he was obtaining a privileged interest on the basis of agreement with his debtor needed to be alert to the possibility that the property concerned might be subject to a legal *hypotheca* which would take precedence over his conventional one. As Nicholas[76] puts it, no creditor could be wholly secure 'unless he knew almost every detail of his debtor's life'.

(e) Other lesser rights
(J.3.24.3)

Emphyteusis. This was a form of long-term or perpetual tenure which probably originated in grants of public land made to private individuals by the state. It came to be used by municipalities; the annual payment of rent maintained the holder of a grant in a position of virtual ownership.[77] Given that the property had to be restored if the interest ended, there was no question of an unrestricted right of disposal. However, the holder's tenancy right was alienable subject, in Justinian's time, to the owner's right of pre-emption or a levy of two percent on the price.[78] On the holder's death the right passed to his successors, but it ended if he died without any (G.3.145; Paul, D.6.3.1 pr). The holder had proprietary remedies; Paul (D.6.3.1.1) refers to an action *in rem* available even against the land-owning municipality. It was a long-standing issue whether *emphyteusis* belonged under sale or hire (see G.3.145); all in all one can see that *emphyteusis* 'no longer leans towards sale or hire but stands apart, with its own implied terms' (J.3.24.3).

Superficies. On the basis of the principle of accession (see section III(g) above) a permanent and fixed structure acceded to the land as a matter of accessory acceding to principal; hence the maxim *superficies solo cedit*. There could be no transfer of ownership of a fixed building as property separate from the land on which it stood and a provision to this effect would be ineffective (Paul, D.44.7.44.1). But a real right in a building, or a defined part of a building, could be held on the basis of a *ius in re aliena*.

To be compatible with the dogmatics of property, the right of *superficies* had to be a lesser one than ownership but, in practice, a holder of the right was in as good a position as an owner. In addition to the obvious rights to occupation and use, the holder could freely transfer his interest, which passed as part of his estate on his death (Ulpian, D.43.18.1.7). That the right was a lesser form of ownership rather than a servitude seems to follow from the fact that it could be subject to a servitude (see Ulpian, D.43.18.1.9).[79] The holder could pass a real security interest in the right. The various remedies applicable to the protection of the right of ownership applied to the protection of the right of *superficies* (see generally Ulpian, D.43.18.1).

Select Bibliography

Birks, P., 'The Roman Law Concept of *dominium* and the Idea of Absolute Ownership', *AJ* (1985) pp. 1-37.

Buckland, W.W. and McNair, A.D., *Roman Law and Common Law* 2nd ed. rev. F.H. Lawson (Cambridge, 1952; repr. 1965) pp. 60-142.

Evans-Jones, R. and MacCormack, G.D., '*Iusta causa traditionis*', in P. Birks (ed), *New Perspectives in the Roman Law of Property: Essays for Barry Nicholas* (Oxford, 1989) pp. 99-109.

Feenstra, R., '*Dominium* and *ius in re aliena*: The Origins of a Civil Law Distinction', in P. Birks (ed), *New Perspectives in the Roman Law of Property: Essays for Barry Nicholas* (Oxford, 1989) pp. 111-22.

Gordon, W.M., *Studies in the Transfer of Property by Traditio* (Aberdeen, 1970).

———, 'The Importance of the *iusta causa* of *traditio*', in P. Birks (ed), *New Perspectives in the Roman Law of Property: Essays for Barry Nicholas* (Oxford, 1989) pp. 123-35.

Johnston, D., 'Successive Rights and Successful Remedies: Life Interests in Roman Law', in P. Birks (ed), *New Perspectives in the Roman Law of Property: Essays for Barry Nicholas* (Oxford, 1989) pp. 153-67.

Pugsley, D., *The Roman Law of Property and Obligations* (Cape Town, 1972).

Rodger, A., *Owners and Neighbours in Roman Law* (Oxford, 1972).

———, 'The Position of *aquae ductus* in the Praetor's Edict', in P. Birks (ed), *New Perspectives in the Roman Law of Property: Essays for Barry Nicholas* (Oxford, 1989) pp. 177-84.

Watson, A., *The Law of Property in the Later Roman Republic* (Oxford, 1968).

Zimmermann, R., *The Law of Obligations: Roman Foundations of the Civilian Tradition* (Oxford, 1996) pp. 220-9.

Notes

1. The inconsistency in this dichotomy is an unavoidable one. While all things having a separate physical existence are potentially property this is by no means true of rights. The latter are not potentially property by their nature but can only be property on the basis of meeting a certain criterion. See section I(b).

2. G.2.12 - 14 and J.2.2.1 - 2 distinguish corporeal and incorporeal things but without noting the importance of identifying the two contexts in which the distinction may apply.

3. Nicholas, *Introduction* p. 98.

4. Id. pp. 99-100: 'the difference between owning and being owed is expressed by the Roman lawyers in the distinction between actions *in rem* and actions *in personam*.'

5. The very word property, in its abstract legal sense, connotes ownership or some other right which accords the holder at least a general position of relative superiority over other claimants – a position protected by an action *in rem*.

6. It did so, as Nicholas, *Introduction* p. 105 notes, 'both because of its intrinsic importance and because of the obvious fact that it is incapable of being moved'.

7. The English law distinction between real and personal property and the Scots law distinction between heritable and moveable property derive from the central importance of land in the feudal scheme of things. In both systems the difference between land and moveable property has greater importance than in Roman law.

8. See P. Birks, 'The Roman Law Concept of *dominium* and the Idea of Absolute Ownership', *AJ* (1985) p. 1.

9. R.W. Lee, *The Elements of Roman Law* 4th ed. (London, 1956) para. 152.

10. See W.W. Buckland and A.D. McNair, *Roman Law and Common Law* 2nd ed. rev. F.H. Lawson (Cambridge, 1952; repr. 1965) p. 65.

11. The translation of T. Mommsen, P. Krueger, and A. Watson (eds), *The Digest of Justinian* IV (Philadelphia, 1985) p. 507; see Nicholas, *Introduction* p. 107.

12. In some circumstances the possessor in the course of acquiring the right of ownership was protected as if he were owner; see end of section III(e).

13. See generally Nicholas, *Introduction* pp. 110-12. One exception was the pledgor who had recourse to the possessory interdict although he did not hold as owner. See section IV(d).

14. It may be noted that the so-called natural law bases are always subject to possible abrogation or modification by the rules of a particular system. In Scots law, for example, feudal notions limit acquisition by an occupier of things not previously owned. See D.L. Carey Miller, *Corporeal Moveables in Scots Law* (Edinburgh, 1991) para. 2.02.

15. Original acquisition was clearly the original form of acquisition and, in this sense, it may be seen to be the primary basis recognised by natural law.

16. This order of treatment departs from the *Institutes*, in which the division is between acquisition by natural law followed by acquisition by civil law; see J.2.1.11. On the question of appropriate order see Lee, *Elements* para. 156.

17. Lee, *Elements* para. 164.

18. Von Savigny identified a 'dinglicher Vertrag' or 'real agreement' as the basis of the passing of ownership. See D.L. Carey Miller, *The Acquisition and Protection of Ownership* (Cape Town, 1986) para. 9.2.2.3(b).

19. Lee, *Elements* para. 158.

20. Ibid.

21. Gaius' statement (G.2.29) that urban praedial servitudes could only be surrendered *in iure* probably followed from the recognition of the greater general importance of servitudes in the urban context.

22. Both *mancipatio* and *in iure cessio* became obsolete in later law and neither are mentioned by Justinian.

23. On this text see W.M. Gordon, *Studies in the Transfer of Property by Traditio* (Aberdeen, 1970) pp. 3-4. As the learned writer points out 'it does not follow that because it is natural that delivery should transfer ownership it is also natural that delivery should be required to transfer ownership'.

24. Ulpian, D.41.1.20 pr, addresses the point directly: 'Delivery should not and cannot transfer to the transferee any greater title than resides in the transferor.

Hence, if someone conveys land of which he is owner, he transfers his title; if he does not have ownership, he conveys nothing to the recipient.' (The translation of A. Watson, et al. (eds), *The Digest of Justinian* IV p. 493.) The first part of this quotation states the principle which has become a basic maxim of property: *nemo dat quod non habet*.

25. Paul, D.41.1.31 pr, says that bare delivery does not transfer ownership but only if sale or some other motivating cause precedes the act of delivery.

26. See R. Evans-Jones and G.D. MacCormack, '*Iusta causa traditionis*', and W.M. Gordon, 'The Importance of the *iusta causa* of *traditio*', in P. Birks (ed), *New Perspectives in the Roman Law of Property: Essays for Barry Nicholas* (Oxford, 1989) pp. 99-109 and 123-35, respectively.

27. See e.g. Julian, D.41.1.36: 'if I give you coined money as a gift and you receive it as a loan, it is settled law that the fact that we disagree on the grounds of delivery and acceptance is no barrier to the transfer of ownership to you.' (The translation of A. Watson et al. (eds), *The Digest of Justinian* IV p. 496.)

28. J.2.1.40 comes close to an outright endorsement of a consensual basis of the transfer of ownership.

29. Gordon, *Studies in the Transfer of Property by Traditio* p. 4.

30. Id. p. 44.

31. Papinian (D.18.1.74) requires the keys to be handed at the warehouse for the device to be effective. Gordon, *Studies in the Transfer of Property by Traditio* p. 59, comments: 'the requirement of presence at the repository suggests that in classical law at least the main function of the delivery of the keys was to give access to the repository.'

32. See Gordon, *Studies in the Transfer of Property by Traditio* p. 14. Where the goods concerned are already held by a third party, there is a case for recognising the transfer of ownership based upon an intimation by the transferor to the third party that he now holds for and on behalf of the transferee.

33. See generally id. pp. 13ff.

34. J.2.6.4 refers to 'a recipient who is in good faith'. See also G.2.43: 'provided that we receive them in good faith.'

35. But without affecting its scope in respect of the acquisition of land; see J.2.6.7.

36. The case of acquisition of enemy property, although more a matter of public than private law, is given as an instance of occupation in J.2.1.17. See A. Watson, *The Law of Property in the Later Roman Republic* (Oxford, 1968) pp. 63ff.

37. Of course, domestic fowl, even though living out and free to come and go, are not naturally wild and remain owned unless abandoned. See J.2.1.16.

38. This text was successfully pleaded by the author in a modern criminal case: *S.* v. *Mnomiya* 1970(1) *South African Law Reports* 66(N).

39. See section II(a) above and Buckland et al., *Roman Law and Common Law*, loc. cit. above note 10.

40. See e.g. the decision of a seven-judge court in *Scottish Discount Co. Ltd.* v. *Blin* 1985 *Session Cases* 216.

41. Lee, *Elements* para. 196.

42. But cf. Nicholas, *Introduction* p. 135.

43. See Buckland, *Textbook* p. 212n15.

44. Nicholas, *Introduction* p. 134.

45. See generally id. pp. 135-6.

46. Justinian says 'He contributes not only his own work but also even part of the material' (J.2.1.25). As Lee, *Elements* p. 142n7, observes, Justinian's primary solution makes the point a superfluous one.

47. 'Much the same rules apply for tenants.' *Emphyteusis* was a form of tenure so close to ownership that acquisition of fruits occurred on separation. See section IV(e).

48. Nicholas, *Introduction* p. 138.

49. See Lee, *Elements* para. 201: 'This mode of acquisition, if it can be called so ....'

50. See R. Feenstra, '*Dominium* and *ius in re aliena*: The Origins of a Civil Law Distinction', in P. Birks (ed), *New Perspectives in the Roman Law of Property: Essays for Barry Nicholas* (Oxford, 1989) pp. 111ff.

51. 'It is a right to a corporeal thing. It must end if its corporeal object ceases to exists' (J.2.4 pr).

52. See Buckland, *Textbook* p. 269. Cf. the case of a plant nursery: Ulpian, D.7.1.9.6.

53. As a *bonus paterfamilias*; see Ulpian, D.7.1.9.2.

54. Nicholas, *Introduction* pp. 145-6.

55. Regarding usufruct and *fideicommissum* see D. Johnston, 'Successive Rights and Successful Remedies: Life Interests in Roman Law', in P. Birks (ed), *New Perspectives in the Roman Law of Property: Essays for Barry Nicholas* (Oxford, 1989) pp. 153ff. Regarding *fideicommissum* see D. Johnston, *The Roman Law of Trusts* (Oxford, 1988).

56. This terminology followed from the tendency to describe land subject to a servitude as *praedium quod servit*. See Buckland, *Textbook* p. 259 (citing Paul, D.8.2.30; Julian, D.h.t.32).

57. See Nicholas, *Introduction* p. 142: '... a right *in rem* over a definite plot of land or a building and annexed to another plot.'

58. See A. Rodger, 'The Position of *aquae ductus* in the Praetor's Edict', in P. Birks (ed), *New Perspectives in the Roman Law of Property: Essays for Barry Nicholas* (Oxford, 1989) pp. 177ff.

59. Buckland, *Textbook* p. 263.

60. See Buckland et al., *Roman Law and Common Law* p. 134: 'The distinction between easements and profits *à prendre* has no place in the Roman Law.'

61. Buckland, *Textbook* p. 259.

62. Cf. ibid.: 'a third party could not be made to repair the wall.'

63. See Lee, *Elements* para. 239.

64. See id. para. 242(4).

65. See J.2.6 pr and section III(e).

66. One would not expect renunciation to require any greater formality than creation.

67. See Nicholas, *Introduction* pp. 150-9 for a comparison of security in Roman and in modern law.

68. See Lee, *Elements* para. 258.

69. See Buckland, *Textbook* p. 431.

70. See id. p. 474 where it is suggested that the use of *mancipatio* may have been retained for some time for the purpose of transfer in security.

71. F. de Zulueta, *The Institutes of Gaius* II (Oxford, 1953) p. 73.

72. As Lee, *Elements* para. 260 points out, this is why *pignus* is dealt with in the *Institutes* (J.3.14.4) as a matter of obligation.

73. J.B. Moyle, *Imperatoris Iustiniani Institutiones* 5th ed. (Oxford, 1912) p. 328.

74. Ibid.

75. Nicholas, *Introduction* p. 152.

76. Id. p. 153.

77. G.3.145: '... so long as the ground rent is paid, it will not be taken away either from the lessee himself or from his heir.'

78. See Buckland, *Textbook* p. 275 (citing C.4.66).

79. Regarding the rule that there could not be a servitude over a servitude, see Paul, D.33.2.1.

4. Succession

William M. Gordon

I. Introduction
(J.2.9.6)

(a) Universal succession

In the institutional scheme J.2.9.6 marks the transition from acquisition of individual things to what is described as acquisition of property in its entirety, acquisition *per universitatem*. The principal case is succession on death, which is what is normally meant by the law of succession in modern law, but in the *Institutes* acquisition of property in its entirety is a wider category. It includes all cases in which one person takes over a whole mass of property (including related obligations) and so succeeds to the previous holder of that mass of property as his universal successor. In contrast, the acquirer of an individual thing may be described as the singular successor of the holder, with the implication that he stands in the same legal position in relation to the thing as the previous holder. The idea of singular succession was certainly familiar to the Byzantine lawyers and has become familiar to modern lawyers through the *Corpus Iuris Civilis*. It has been argued that all the texts in which singular succession appears are interpolated, but the argument for such wholesale interpolation is not convincing.[1] What can be said is that succession referred primarily to universal succession, as in J.2.10.11, and some texts have been generalised to refer also to singular succession.

The cases of universal succession mentioned in J.2.9.6 are acquisition by the heir of a deceased person (J.2.10 - 3.8), acquisition by an estate-possessor (J.3.9), acquisition by an adrogator (J.3.10), and acquisition by an assignee to preserve freedom (J.3.11). Acquisition of the estate of a debtor on sale of his assets and of the estate of a free woman enslaved under the Claudian Resolution are referred to at the end of this run of titles (J.3.12), but they are referred to only to be dismissed, the former as obsolete and the latter as abolished by Justinian himself. Other obsolete cases which no longer even warrant mention in the *Institutes* are acquisition of an independent woman's property when she passed into marital subordination (G.3.82 - 84; see Chapter 2, sections II(d) and III(b) above) and assignment in court of a statutory inheritance (G.3.85 - 87). Acquisition by the adrogator and assignee to preserve freedom are treated very briefly. The title on the estate-possessor is also brief, but there are numerous references to estate-possession in other titles which discuss praetorian modifications of the state law, which regulated the position of the heir. It is clear that the position of the heir, and succession on death, are central to the treatment.

(b) The heir

Acquisition of property in its entirety assumes acquisition of assets; J.2.9.6 says *res* ('assets') without further qualification. In J.3.10.1 (and G.3.83) the expression used is 'corporeal and incorporeal assets and claims', apparently treating claims as falling outside the class of incorporeal assets, although in J.2.2.2 they are included in that class. J.2.2.2 also includes an inheritance itself among incorporeal assets, whereas J.2.9.6 refers to the assets within an inheritance as what is acquired when one becomes an heir. It is important to note that whether an inheritance is regarded as itself a *res* or as the basis of acquisition of other *res*, the heir who acquires it acquires not merely assets but liabilities. In a number of texts, such as D.50.17.62 (Julian), it is said that the heir succeeds to the whole legal position of the deceased, meaning that he is universal successor not only in the sense of taking over the entirety of the property of the deceased, but in the sense that he succeeds to and hence represents the deceased in his legal relationships. In principle, as he takes over from the deceased, he is therefore liable for all the deceased's debts; only Justinian made it the normal rule that an heir's liability was limited to the assets of the estate (see J.2.19.6), although the heir was protected from creditors in certain circumstances before Justinian (see section II(f) below). As the heir was the deceased's successor, an heir could not be appointed to act for a limited period – once an heir always an heir (*semel heres semper heres*; see Gaius, D.28.5.89(88); J.2.14.9) – but this principle was breached in fact although not in theory by the institution of trusts to hand over an inheritance (J.2.23; see section II(h) below).

There were qualifications to the principle that the heir took over from the deceased. He or she did not succeed to the deceased's position in his family as husband or as wife. A male heir did not become head of the deceased's family[2] (although in the earlier law the heir was primarily responsible for maintenance of the *sacra*, the family cult; see G.2.55[3]). The heir did not succeed to any public office held, including the office of guardianship, and to rights such as usufruct which were limited to the lifetime of the holder (J.2.4.3), or to rights which were personal to the deceased, such as the right to sue for contempt (J.4.12.1). On the other hand, some liabilities died with the deceased, for example delictual liability (in principle, J.4.12.1), and some contractual liabilities, such as that of personal surety or surety in classical law (G.3.120), or liabilities which were regarded as personal to the contracting party, such as that of a partner (J.3.25.5) or mandatary (J.3.26.10) (although the heir had to complete unfinished business which the deceased had begun). A further limitation is that he succeeded to the legal position of the deceased and not to the factual position, so that, for example, he did not obtain possession of property by becoming heir, although he became owner.[4] The heir on the other hand might come under new liabilities imposed on him by the deceased, for example, to carry out legacies or trusts. Joint heirs, in principle, shared rights and duties relating to the inheritance (see section II(d) below).

(c) The heir as executor

In modern Scots and English law the function of winding up an estate and carrying out the wishes of the deceased is performed by an executor, appointed by the deceased or by law, and acting under authority of a court. He need have no interest in the estate and, as executor, carries out an administrative function, but he succeeds to the rights and liabilities of the deceased, with his own liability restricted, in principle, to the value of the estate which he administers. When the Roman heir's liability comes to be similarly restricted to the value of the assets which he inherits, when he need not obtain any benefit from the estate (see C.6.30.22 of AD 531 and section II(h) below), and when he has lost any responsibility for the family cult, so that he is concerned solely with the continuance of patrimonial rights and liabilities, he bears quite a close resemblance to the executor whose functions he also performs, but it is a resemblance in function. The heir is in principle a successor to the deceased, not merely an administrator of his estate.

(d) Testate and intestate succession

As explained in J.2.9.6, an inheritance may devolve either under a will or under the rules governing succession in the absence of a will. The rule that it is impossible for anyone to die partly testate and partly intestate (J.2.14.5) meant that in Roman law testate and intestate succession were regarded as mutually exclusive; one could not leave part of one's estate by will and let the rest devolve under the rules of intestate succession. For the early law at least, when the purpose of a will may have been simply to appoint a successor, whether from among or in place of the relatives who would otherwise succeed, the rule is logical, but as soon as it becomes possible to appoint joint heirs and to leave legacies it ceases to be so. This position may have been reached by the time of the *Twelve Tables*, but the rule remains formally intact even in Justinian's law, except for soldiers (J.2.14.5; Ulpian, D.29.1.6), who from the late Republic and early Empire enjoyed various privileges in the matter of making wills (J.2.11; D.29.1). However, the rule could in fact be breached by imposing trusts on one's heir on intestacy, even a trust to convey away the whole estate (J.2.23.10); in theory the rule stood because the codicil by which such a trust could be imposed was not technically a will (J.2.25.1) and so the deceased was still intestate. A will could also be left standing in part if, for example, a challenge that it was irresponsible (J.2.18) was successfully made against only one of two or more heirs.

It has been argued that, historically, in Rome testate succession preceded intestate succession in the sense that the head of a family nominated his successor.[5] However, our earliest evidence, the *Twelve Tables*, appears to recognise both types of succession – see *XII Tab*. V.3, 4 (*Roman Statutes* II no. 40, pp. 638, 641) – and so it is difficult to settle the question conclusively, even accepting that the 'will' referred to in the *Twelve Tables* was something different from the will as known in later law. It seems inherently more probable that there was some customary rule of disposal of property on death before any form of will became possible. It is also quite probable that some property such as land or other 'family' property could not be disposed of by will when wills were first recognised, although in the developed law there was complete freedom of testation in this respect. Until the appearance of the complaint of an irresponsible will (J.2.18), indeed, there was complete freedom of testation provided that the formal rules on disinheriting children (J.2.13) were observed.

Succession on death is a common subject of dispute or discussion in the texts,[6] and the texts on testate succession outnumber those on intestate succession. Almost twice as much of the *Institutes* is devoted to the former as to the latter. This reflects the greater complexity of the law of testate succession and has little bearing on the question whether the Romans preferred testate to intestate succession. Maine's reference to 'the singular horror of Intestacy which always characterised the Roman' is an exaggeration,[7] but it would also be wrong to assume that only wealthy Romans made wills or went to law over questions of succession. There is evidence that Romans of modest means left wills;[8] succession is a fertile source of family disputes and there was, in theory at least, easy access to jurists or the emperor for advice on the legal questions arising. It may be added that succession was a matter of importance in the Roman family and there is considerable discussion of succession in the literature on Roman social and family life.

II. Testate Succession on Death
(J.2.10 - 25)

Titles 10 to 25 of book 2 deal with testate succession. Titles 2.10 and 2.11 start by dealing with the types of will which are found and what might be described as the 'external' formalities, such as the number of witnesses required. Title 2.12 then deals with capacity to make a will, a question which might logically have come first (see G.2.114) but which can quite practically be postponed, because unless there is what appears to be a will it is unnecessary to discuss whether it is valid as an expression of the purported testator's intentions. Title 2.13 begins what can be regarded as the 'internal' requirements of a will by discussing which descendants need to be disinherited. Titles 2.14 to 16 consider the positive requirement of institution of an heir and substitution of alternative heirs if the first heir does not take. Invalidation of a will and challenge to it as irresponsible are dealt with in J.2.17 and 18. If a will is valid it must be put into effect, and J.2.19 explains how the various types of heir obtain the inheritance with its rights and duties. The following two groups of titles, J.2.20 to 22 and J.2.23 to 24, deal with the burden imposed on an heir by legacies and trusts respectively, and the last title, appropriately enough, deals with codicils. There is no systematic account of manumission of slaves by will (J.1.5.1; 1.6.7; 1.7), appointment of testamentary guardians (J.1.14), or appointment of supervisors (J.1.23.1, 5) (on which see Chapter 2, sections I(c); V(a), (e) above), because the standpoint is the acquisition of property in its entirety; in J.2.20 pr there is an apology for dealing with legacies out of the context of acquisition of individual things, with the reasonable excuse that they fit well enough into the context of testamentary acquisition.

(a) Types of will
(J.2.10 - 11)

The etymology suggested for *testamentum* in J.2.10 pr (*testatio mentis*, 'testimony to the mind') is fanciful and in this respect typical of other etymologies in the texts.[9] It seems to go back at least to Servius Sulpicius, as we learn from Aulus Gellius, *Noctes Atticae*, 7.12.1. The suffix *-mentum* is not derived from *mens* ('mind'), but refers to instrumentality in a concrete sense, so that the will is the means by which the deceased makes a declaration of his wishes before witnesses.[10]

The brief historical account in J.2.10.1 - 2 can be supplemented from G.2.101 - 108. Gaius adds little to our scanty knowledge of the will made before the convocation and the will made in battle-line, both of which pre-date the *Twelve Tables* but were obsolete by the second century BC. He does give fuller details of the procedure for making a will by bronze and scales, which was the will valid by state law in his day. This type of will was apparently a product of the pontifical jurisprudence adapting mancipation (G.1.119 - 122) for the purpose. According to Gaius (G.2.103), the property-purchaser originally took a conveyance of the estate and carried out the testator's wishes, but by his time the mancipation to the property-purchaser was purely formal. The testator appointed the heir in a document to which he referred in the course of the ceremony (G.2.104) but the contents of which were not disclosed. The heir thus appointed was responsible for carrying out the testator's wishes.

Normally the document referred to was subscribed and sealed by the testator and by the seven other participants in the ceremony – the witnesses, the holder of the scales, and the property-purchaser – and the praetorian 'will' referred to in J.2.10.2 might well be such a document.

However, a praetorian will took effect by the praetor's grant of estate-possession in support of it, and the praetor would grant estate-possession if an appropriate document were produced, even if the ceremony had not taken place (G.2.119). There has been scholarly debate on whether a praetorian will of this sort was a will or not.[11] Strictly speaking it was not, because in the absence of the formal ceremony the heir named would not be heir by state law; he would simply be an estate-possessor. If there were heirs on intestacy with a right under state law they could recover the estate from him and his estate-possession was described technically as provisional (*sine re*) (see section III(c) below). However, the emperor Antoninus Pius (AD 138-61) allowed the estate-possessor to defeat any claim brought by intestate heirs by using the defence of fraud (*exceptio doli*, G.2.120 - 121, 149a). His estate-possession was then *cum re* (see section III(c) below) and at least thereafter it seems right to speak of a praetorian will, because the successor appointed by it was in virtually the same position as an heir by state law. Gaius leaves unclear whether such a will was effective if made by a woman under guardianship (see Chapter 2, section V(f) above) without the endorsement of her guardian (G.2.121), but the implication appears to be that it was effective if the endorsement could be compelled (G.2.122).

The 'three-fold' will of J.2.10.3 appears in C.6.23.21, a pronouncement of the emperors Theodosius II (AD 408-50) and Valentinian III (AD 425-55) of AD 439. The additional requirement imposed by Justinian (J.2.10.4) was withdrawn in AD 544 by Nov.119.9 because it was creating too many problems. It is not clear how far his fears about forgery or fraud were justified.

Testamenti factio (J.2.10.6) can refer either to capacity to make a will (when it is said to be active) or to capacity to take under a will (when it is said to be passive); according to Papinian (D.28.1.3) it is a matter of public law. In this context in the *Institutes* it may refer to capacity to take, as J.2.10.8 - 11 contemplates persons under authority (*potestas*) acting as witnesses, and these persons could not make a will, unless they were soldiers dealing with their military fund (J.2.12 pr). The exclusion of women as witnesses reflects a view that they could not take part in public acts. For the other exclusions compare J.2.12.1 - 3. Among the infamous persons excluded were adulterers (see Papinian, D.22.5.14) and there are specific provisions in later law excluding heretics (C.1.5.4 of AD 407) and apostates (C.1.7.4 of AD 426). The written reply of the emperor Hadrian (AD 117-38) is preserved in C.6.23.1 (no date), to the effect that it was not important if certain witnesses turned out to be slaves, since they had served as witnesses without objection by those concerned. A similar rule, expressed in the maxim *communis error facit ius* (literally, 'a shared error makes law'), is attested in other contexts, notably in Ulpian (D.1.14.3, a fugitive slave acting as praetor).

A provision initially valid might be forfeited for unworthiness (*indigni-*

tas) where the beneficiary acted in what was regarded as a reprehensible fashion, as where a testamentary guardian who also benefited from the will appointing him claimed to be excused from undertaking the guardianship (Paul, D.34.9.5.2). There are many other examples, as well as discussion of when conduct was to be regarded as unworthy, in D.34.9 and C.6.35. Commonly forfeiture was to the imperial treasury.

Paragraphs 9 to 11 of J.2.10 may be compared with G.2.105 - 108 and the adjustments made by Justinian noted. They result mainly from the abandonment of the mancipatory will, which Gaius describes as a transaction between testator and property-purchaser (G.2.105). There is no general rule that beneficiaries may not act as witnesses (J.2.10.11); only the heir and members of his family are excluded, on the ground that the heir becomes successor of the testator, and not simply on the ground that he is a beneficiary. Gaius' recommendation not to use the heir or a person in his *potestas* as a witness to a mancipatory will (G.2.108; J.2.10.10) seems on the other hand to be based on the inadvisability of using a person with an interest in the estate.

Other forms of will receive passing mention in the *Institutes*. The oral or nuncupative will referred to in J.2.10.14 is regulated by Justinian's pronouncement in C.6.23.26 (AD 528). Paragraph 2.12.4 refers to the special provision for blind testators made by the emperor Justin (AD 518-27) in C.6.22.8 (AD 521), involving use of a notary or an eighth witness to record the will. Other forms of will not mentioned in the *Institutes* are (1) a will recorded in the archives (C.6.23.18 of AD 397); (2) a will recorded with the Emperor (C.6.23.19 of AD 413); and (3) an informal will distributing property among one's descendants (C.6.23.21.3 - 3a of AD 439) (part of the pronouncement regulating the normal three-fold will described in s.3) as amended by Nov.18.7 (AD 536) and Nov.107 (AD 541), but (4) Justinian in C.6.23.21 (AD 534) also makes concessions on subscription by the seven witnesses in country places where there is a dearth of literate people and allows a minimum of five witnesses if seven cannot be found. The informal will distributing property among descendants might be holograph but Justinian does not recognise the general holograph will allowed by Valentinian's Nov.21.2 (AD 446), which remained confined to the West.

Military wills (J.2.11). There appears to be no direct historical connection between the will in battle-line (J.2.10.1) and the military will. According to Ulpian (D.29.1.1 pr) the latter originated in a temporary concession by Julius Caesar (d. 44 BC) and successive grants of the privilege thereafter by the emperors Titus (AD 79-81), Domitian (AD 81-96), Nerva (AD 96-8), and Trajan (AD 98-117, whose written reply is quoted in J.2.11.1 and by Florentinus, D.29.1.24) before it became a standing concession by imperial pronouncement. It is suggested that the extension of the privilege is connected with the narrowing of the scope of trusts, which originally could benefit non-citizens but were then restricted (G.2.285).[12] The privilege of making an informal military will is distinct from the privilege granted to soldiers still under *potestas* of making a will of their military fund, referred to in J.2.12 pr, but a military will could be used to dispose of the military fund in appropriate circumstances (Tertullian, D.29.1.33). In the case of the quasi-military fund, referred to in J.2.11.6, when a will could be made (a privilege generalised by Justinian, see C.3.28.37.1 of AD 531) the will had always to conform to the general law, although it could not be challenged as irresponsible while the holder of the quasi-military fund was under *potestas* (C.3.28.37.2 of AD 531). For other privileges of soldiers, see J.2.11.3 - 4 and the discussion below.

The informality of a military will could give rise to difficulty of proof of testamentary intention. J.2.11.1 might seem to imply that witnesses were necessary – 'if he collected people together' – but the point is rather that there must be proof of immediate intention to make a will as opposed to indication of a possible desire to benefit someone. An imperial pronouncement (C.6.21.15.1 of AD 334) speaks of a will written in blood on the soldier's shield or the sheath of his sword, or with his sword in the dust of the battle-field, as valid.

As the occasion for discussion of the military will is remission of the normal formalities, J.2.11 does not give a full account of the other privileges enjoyed by soldiers in relation to testation. Only two are mentioned: the fact that a deaf and dumb soldier could make a will (J.2.11.2, without taking advantage of the concessions referred to in J.2.12.3), and that status-loss by adrogation did not invalidate a military will (J.2.11.5); the first of these was of limited value as a deaf and dumb soldier should be discharged as unfit (Ulpian, D.29.1.4). Other privileges were: (1) he could leave a will of part of his estate (J.2.14.5; Ulpian, D.29.1.6); (2) he could appoint an heir for a limited period (C.6.21.8 of AD 238) or from a specified date (Tryphoninus, D.29.1.41 pr); (3) he could provide a substitute for an heir who accepted the inheritance (Ulpian, D.29.1.6) and for an emancipated son in respect of what he left them (Tryphoninus, D.29.1.41.4) and for a ward, even if he made no will or his own will did not take effect (Tryphoninus, D.29.1.41.5) (for the general rules see J.2.16); (4) he did not require a specific exclusion to disinherit children and he could disinherit them by implication (J.2.13.6; Ulpian, D.29.1.7; C.6.21.9 of AD 238); (5) there could be no complaint that his will was irresponsible (Ulpian, D.5.2.8.3 - 4; C.6.3.28.9 of AD 223); (6) the Falcidian Act (J.2.22) did not apply (C.6.21.12 of AD 246); (7) he could appoint as heirs or legatees persons who normally were disqualified from accepting or from taking benefit, such as foreigners and (in the classical law) Latins and the unmarried and childless (G.2.110 - 111); and (8) he could appoint an heir by codicil (Papinian, D.29.1.36 pr).

There were, however, limitations on the privileges enjoyed. The Aelian-Sentian Act (J.1.6) applied, and the Fufian-Caninian Act (see Chapter 2, section I(c) above) until its repeal by Justinian (J.1.7), as did the Julian Act on dotal property (Paul, D.29.1.16). A soldier could not make a will while captive (Ulpian, D.29.1.10) and so enslaved (J.1.3.4), nor could he leave legacies by way of penalty until Justinian made a change in the general law (J.2.20.36). The privilege began at earliest on enrolment with the colours (Ulpian, D.29.1.42); its application during service and on discharge is explained in J.2.11 pr, 3 - 4. As the privilege was said to be granted because soldiers were unskilled in legal affairs, it applied only to the lower ranks (Ulpian, D.29.1.44). Any soldier could have a military fund, however.

(b) Capacity to make a will
(J.2.12)

Capacity to make a will is treated in a negative way by listing those incapable of making a will (J.2.12). The basic rule that only a Roman citizen might do so is taken for granted; Gaius probably mentioned foreigners, Latins, and capitulated aliens (see Ulpian, *Rules* 20.14) although most of the passage in which he seems to have dealt with capacity is lost (see G.2.111). The categories of Latins and capitulated aliens had gone under Justinian (J.1.5.3) and the extension of citizenship made it unnecessary to mention foreigners. In each of the cases mentioned it is stressed that the capacity must exist at the time the will is made. Although the will takes effect only on death, the act of making it is *inter vivos* and must be validly performed at the time when it is done. This may, but need not, reflect the derivation of the will by bronze and scales from conveyance by mancipation, which normally would take immediate effect.

The inability of persons under authority to make a will follows from the fact that they had nothing to dispose of until the military fund was introduced by Augustus for soldiers, and testamentary capacity in respect of the fund was given as described in J.2.12 pr. When the quasi-military fund was introduced in the post-classical period only some holders were given testamentary capacity in respect of it, but Justinian granted such capacity to all (J.2.11.6). The further extension of proprietary capacity described in J.2.9.1 was not accompanied by power to make a will in respect of the property concerned (see C.6.60 and 61), but imperial donations were equated to the military fund by Justinian (C.6.61.7 of AD 530).

In J.2.12.3 'the dumb and the deaf' must be read as one phrase, but a will by bronze and scales could not be made by a person who was either deaf or dumb (Ulpian, *Rules* 20.13 and the ambiguous text, Gaius, D.28.1.6.1) because it was necessary both to speak and to hear what was said by the property-purchaser. But from the pronouncement to which Justinian refers, C.6.22.10 (AD 531), it seems that the question was disputed and there is no obvious reason why a deaf or dumb person could not make a praetorian will. Justinian draws up a scheme which distinguishes between the deaf and the dumb. It provides that natural deaf mutes have no capacity (although they might be allowed to make a will by imperial privilege, see Aemilius Macer, D.28.1.7); the deaf or the dumb may make a written will, and those who have become deaf or dumb or deaf mutes are not deprived of the capacity they had. The pronouncement stresses that no distinction is made between male and female.

The incapacity of captives followed from the fact that they became slaves and slaves in general could not make a will; slaves in the service of the state are an exception (Ulpian, *Rules* 20.16). Slaves are included among the persons under authority referred to in the J.2.12 pr, although it does seem to contemplate free persons by the reference to the permission of parents. The Cornelian Act, dated to around 81 BC, saves a will already made by a person before capture; apart from this provision it would have been invalidated by the status-loss resulting from enslavement (J.2.17.4). The right of rehabilitation restored a will made before captivity if the captive was honourably released.

In Justinian's law women could make wills in the same way as men and had capacity from the age of twelve, not fourteen. In the time of Gaius the position was complicated by the perpetual guardianship of women (on which see Chapter 2, section V(f) above). To make a will valid by state law a woman required the endorsement of a guardian. There are texts which suggest that the guardian in question had to be one created by means of a fiduciary contrived sale (e.g. G.1.115, 115a; 2.112 - 113) until Hadrian abolished the requirement of a contrived sale, leaving the rule that a woman who did have a guardian still needed his endorsement (G.1.115a; 2.112). Thus a woman who was freed from guardianship, say by having the privilege of children (G.1.194), would need no endorsement. A woman could make a praetorian will which from the time of Antoninus Pius was apparently effective even without a guardian's endorsement, unless the guardian was a parent or patron, because they, as guardians, still had real control over the woman's acts (G.2.122). It is argued that the rule by state law was that a woman with a guardian needed endorsement of her guardian and Hadrian relaxed it, allowing a woman with a guardian to compel endorsement unless the guardian was a parent or patron.[13]

Even if a will complied with all the rules on writing and witnesses etc. and was made by someone with the necessary capacity, it might fail in effect if it were not drawn up correctly or did not take account of the claims of members of the testator's family to benefit. Some of the relevant rules were matters of form, others were matters of substance; they are dealt with in the following group of titles.

(c) Disinheriting
(J.2.13)

The logic of starting with a title on disinheriting children is that a will might be totally invalid or fail to take effect as expected if the rules on disinheriting children were not observed. It was also permissible to start a will by dealing with disherison, even before naming an heir. That it was necessary to disinherit may reflect a stage in the law where the family had an actual or potential interest which could not be overridden by a will, but two things must be noted. On the one hand, the rules gave no substantive protection – no assurance that benefit would be received. That became the function of the rules on irresponsible wills (J.2.18), apart from the special provision for a person adrogated under puberty (J.1.11.3); the rules on disinheriting simply required that notice be taken of those who had to be mentioned. On the other hand, the fact that a child was disinherited did not necessarily mean that he or she would receive no benefit. It meant only that he or she would not receive the benefit as heir. This, at least, was true until Justinian reformed the law by Nov.115 (AD 542) and, among other things, required that children be instituted heir for some share if the will was not to be open to challenge as irresponsible (see section (e) below). Before Nov.115 a disinherited child could, for example, receive what was regarded as due to him or her by legacy or trust.

The first four paragraphs of title 13 summarise the law as it stood before the reforms made at the time when the *Institutes*, *Digest*, and *Code* were being compiled. Much is taken from G.2.123 - 137 from which we learn some further details, for example, that there was a dispute between Sabinians and Proculians on the effect of predecease of a son who was omitted. The Sabinians thought that the will was still a nullity; the Proculians thought that the will was saved because it was not radically null (G.2.123). (The Sabinians also took the view that the live birth of a posthumous child for whom no provision was made broke the will even if the child uttered no cry – a view supported by Justinian: C.6.29.3 of AD 530). We also learn of a restriction on the rights of daughters made by Antoninus Pius (G.2.126); they were never to get more than they got by their right to come in to share with the appointed heir, given to them by state law (the *ius accrescendi*), although the rules on estate-possession might be more generous.

Praetorian law already made innovations on the rules of state law, taking account of emancipated children and increasing the rights of females (until Antoninus Pius in turn restricted them). Justinian then recast the law by C.6.28.4 (AD 531), the provisions of which are briefly given in J.2.13.5. He equated the position of males and females and required express mention of all those to be disinherited. It was no longer enough to use the clause 'let all others be disinherited' as a catch-all provision. Again, his alteration of the law of adoption (J.1.11.2) had the consequential effect that only in the case of full adoption were children taken out of their family and so out of the scope of the rules on disinheriting.

As J.2.13.7 explains, the rules on disinheriting did not apply to mothers or maternal ascendants because they are basically concerned with the agnatic family and succession to the head of the family. A mother was not head of her household in this sense. Independent women would also have been under guardianship in the early law. The complaint of an irresponsible will was available if descendants were omitted, however, and a woman might put a clause disinheriting children in her will simply to show that they had not been overlooked, as we see from Pliny's *Letters*, V.1.

Where a challenge was made to a will under the praetorian rules on disinheriting descendants, it was made by claiming estate-possession counter to the will (J.3.9.3). The grant of such estate-possession did not mean that the will was overruled completely; the will still stood if valid by state law, which it might be if, for example, an emancipated child had been ignored, and it had limited effect. Persons properly disinherited remained disinherited (Ulpian, D.37.4.8 pr); provisions for substitutes for children (J.2.16) were still effective (Africanus, D.28.6.34.2); appointments of testamentary guardians were confirmed by the relevant magistrate (Julian, D.26.3.3), and legacies to ascendants and descendants and legacies of dowry were still payable (although not to the claimant of estate-possession), up to a proportionate share of the estate with the estate-possessor (D.37.5). Subject to this, the estate-possession was in classical law effective against the heir; it was *cum re* (see section III(c) below).

(d) Appointment of heirs and substitute heirs
(J.2.14 - 16)

The valid appointment of an heir was crucial. The appointment of the heir is described by Gaius (G.2.229; cf. J.2.20.34) as the foundation or cornerstone of the will; keystone might be an even better term because without it the will would collapse. Gaius (G.2.116 - 117) stresses the importance of the correct imperative form; he remarks that even the words 'I institute X as my heir' or 'I make X my heir' were generally regarded as unacceptable, and this although the appointment of an heir is technically referred to as his 'institution'. The argument for such exactitude is that it should be beyond doubt that the testator is making a present appointment and not merely expressing an intention or wish for the future, but Constantine (AD 312-37) or Constantius II (AD 337-61) swept aside such niceties (C.6.23.15 of (?) AD 339), laying down that any expression of intention would suffice. It was also essential in classical law that the appointment of the heir should be the first substantive provision; nothing which diminished his rights could effectively precede the appointment (G.2.229 - 231). Justinian abolished this rule (J.2.20.34; C.6.23.24 of AD 528). He therefore proceeds directly to discussion of who could be appointed.

That slaves could be appointed heir might seem somewhat surprising, although the appointment of one's own slave did ensure that there would be an heir (J.2.14.1; 2.19.1). The slave could not benefit personally unless he were given freedom. Gaius states without qualification that freedom must be specifically given (G.2.186), but it is not surprising to find Justinian quoting a view of Paul that a grant of freedom could be implied from appointment as heir. Justinian's own elaborate pronouncement (C.6.27.5 of AD 531) is to the effect that a grant of freedom can be assumed if this appears from the terms of the will; appointment as heir suffices, but not the mere grant of a legacy or trust. The express mention that a slave given in usufruct belongs to the testator, and that a slave held in usufruct by the testator belongs to someone else, might imply some uncertainty over the conception of usufruct.[14] Ulpian notes that where a usufruct over a slave has been given to someone else, the appointment does not take effect until the end of the usufruct (D.28.5.9.20). A slave belonging to someone else could not accept the inheritance without specific authorisation (J.2.9.3; 2.14.1, 3) because his master would become heir with all the responsibilities of heir and, in principle, a slave could not under state law impose obligations on his master or under praetorian law expose his master to liabilities without the master's consent.

J.2.14.2 contains two interesting points. The first is the reference to the *hereditas iacens*, an estate awaiting acceptance by an outside heir (see J.2.19 and section (f) below). In this text the inheritance is said to maintain the legal personality of the deceased. The second point is the recognition of a person still in the womb as having legal existence.[15]

J.2.14.4 - 8 describes the peculiar way in which the estate was divided between co-heirs in 'ounces', treating the whole estate as a pound. As already observed (sections I(a) and (b) above) the whole estate had to be distributed, and the interpretations given show a very artificial system, or assume a very precise knowledge of the law on the part of testators, if the division is supposed to be reconcilable with the testator's intention. Where there were co-heirs, debts of the deceased were automatically shared between them, by a provision attributed to the *Twelve Tables* (table V. 9; *Roman Statutes* II no. 40, pp. 648-9), but debts, legacies, or trusts could be charged by the testator on particular heirs. There was a special action of division of the estate (J.4.6.20) in which there could be adjustment of liabilities arising between the co-heirs as a result of their use of the property (J.4.17.4). Obligations arising between them are described as quasi-contractual (J.3.27.4) (see Chapter 5, section VII below).

The rule that the institution of an heir could not be limited by a specific date is in one sense a logical consequence of treating the heir as successor to the deceased. Once the heir has taken up the position of heir, he is the successor: *semel heres, semper heres* ('once heir always heir'). The denial of appointment from a certain time could be explained as a consequence of the use of mancipation in the mancipatory will, but would also be in the interests of creditors and performance of the *sacra*, as there would be no point in deferring taking up the position. Ignoring the time stated and treating the appointment as immediately effective recognises that the deceased did wish to die testate. Although the position of heir could not be temporary, the estate could in fact be passed on after a period by means of a trust (J.2.16.9; 2.23). Conditional appointments were permitted, for example, to provide for the appointment of an alternative heir in the event the first-named heir does not take up the inheritance. (The case of substitutes is a different matter and is dealt with in J.2.15.) Other conditions were also possible. Allowing conditions assumes development from what seems to have been the original form of the mancipatory will, in which conditions would not have been possible.

The rule that impossible conditions are ignored, stated in J.2.14.10, oversimplifies the position, at least for classical law; it is, however, stated equally baldly by Ulpian, D.28.7.1. The text does not explain the meaning of impossibility in this context, nor does it distinguish between initial and supervening impossibility or between impossibility supervening the making of the will and supervening the death, when the will takes effect. Other texts show that impossible means impossible by nature – a typical example is 'if X touches the sky with his finger' – or impossible in fact, for example, a condition of freeing a slave who is already dead. Legal impossibility should also be relevant, but there are no texts referring to institution of heirs in this respect. In classical law there was a view that an institution made under an impossible condition was void. Gaius states that this is the Proculian view on conditions in legacies (G.3.98), and Q. Mucius Scaevola appears to have taken this view on institutions of heirs; other writers may have distinguished between initial and supervening impossibility and regarded impossibility supervening the making of the will but occurring before the death as avoiding the institution.[16] The argument would be that a new institution was possible and the imposition of an impossible condition suggests no serious intention of making an institution. Ignoring an impossible condition, on the other hand, saves the will, which the testator at least purported to make.

Conditions other than impossible conditions might also be ignored. Marcian lists conditions framed against legislative provisions, against sound morals, or derisory or disapproved by the praetor (D.28.7.14).

A condition means something inherently uncertain and not merely something not immediately certain but ascertainable. A condition 'if X was consul' was not conditional in the sense of suspending the operation of the institution, although whether the institution took effect or not depended on whether X had been consul. There could not be a resolutive condition, as the appointment of an heir once effective could not be rescinded. Whether a conditional appointment took effect would normally depend on fulfilment of the condition, but a condition could be treated as fulfilled if the potential heir had done what he could but some other person had failed to provide the co-operation necessary for fulfilment of the condition, for example, had refused to accept payment of a sum of money. The precise scope of this rule, which may be compared with the similar rule on conditional obligations, is unclear.[17]

J.2.14.12 appears to be a remnant of discussion of the classical requirement that the heir named must be a definite person (see G.2.242). The requirement that a beneficiary be a definite person also applied to legacies (see G.2.238 - 242; J.2.20.25 - 29), but Justinian states that he amended it not only for inheritances but for legacies and trusts (J.2.20.27), referring to a pronouncement (C.6.48.1 of AD 528-529) of which only a Greek version survives. On one view he abolished the requirement that a beneficiary be a definite person as previously understood, but this has been questioned.[18] The texts make clear that there was discussion on how definite an idea of the potential beneficiary the testator had to have had in order to validate the provision in the earlier law.

In classical law there was difficulty over the appointment as heirs of bodies such as towns, where the membership was fluctuating, because the appointment was seen as appointment of the members of the body who were unidentified persons (Ulpian, *Rules* 22.5). The doctrine of legal personality of such corporate bodies was not well developed. However, exceptions were made by senate resolution for appointments of towns made by freedmen of the town and by special privilege (Ulpian, *Rules* 24.28). Constantine allowed appointment of the Church (C.1.2.1 of AD 321). There was general provision for appointment of civic bodies (*civitates*) by the emperor Leo (AD 457-74) in AD 469 (C.6.24.12), and under Justinian the difficulty was removed (see C.6.48). He also made special provision to save appointments of 'the poor' or 'captives' with mechanisms to put such appointments into effect, using existing foundations or the Church (C.1.3.48(49) of AD 531).

Ordinary substitutes (J.2.15). The title on ordinary substitutes deals briefly with the common case in which a testator made provision for a series of appointments as heir, each appointment conditional on failure of the preceding appointment to take effect. One reason for the appointment of a slave as the last in the series was to ensure that there was an heir to take over, particularly where there were doubts about the solvency of the estate (see J.2.19). However, such an appointment could be made even when there was no fear of insolvency (see Pliny, *Letters*, IV.10).[19] The *principium* and paragraph 1 of title 2.15 appear to be taken from Marcian's *Institutes*, book 4 (see D.28.6.36 pr - 1), although very similar to G.2.174 - 175. The passage in Gaius shows that where several heirs were named in succession they might be given a limited period in which to decide whether to accept the inheritance, and might be required to make a specific formal act of acceptance called *cretio*, rather than accepting by acting as heir (*pro herede gestio*, see J.2.19.7); the same provision might be made where there was no substitute heir and the choice was between the testamentary heir and the heir on intestacy (see G.2.164 - 173).

J.2.15.2 - 4 deals with difficulties of interpretation in substitutions of which there are many further examples in the relevant *Digest* title (D.28.6). The written reply of the emperors Severus (AD 193-211) and Antoninus (AD 211-17) referred to in J.2.15.3 is not preserved in the *Codex*. (The same rule is given by Julian, D.28.6.27, and Papinian, D.28.6.41 pr.)

Substitutes for children (J.2.16). The unusual provision allowing fathers, in effect, to make wills for their descendants in *potestas* who succeeded to them as testamentary heirs (J.2.16 pr) or who were disinherited by them (J.2.16.4), is discussed by Gaius in a passage (G.2.179 - 184) which corresponds closely to this title. An innovation by Justinian, made in C.6.26.9 (AD 528) and referred to in paragraph 1, is substitution for children or other descendants who were insane and who, like those under age, could not make wills for themselves. In their case, however, there were restrictions on who could be appointed; if the insane descendants had sane children or, failing descendants had sane brothers, only those sane relatives or any one or more of them could be made a substitute. Although the choice of heir is made by the father, the substitute heir is heir to the child, as is made clear by paragraph 4 which points out that all acquisitions made by the child while under age go to the substitute. It is not a case of the father disposing of what he himself has left to his descendant.

The precaution described in J.2.16.3, of concealing the substitute's name until the death of the child, is an obvious one. In describing the act of concealment Gaius (G.2.181) refers to 'tables' where Justinian, at the beginning of the paragraph, refers to 'parts of the will'. Gaius assumes a will written on wooden tablets smeared with wax and tied together, which was the typical Roman document at his time, but the principle of concealing the substitution applies to any form of document.

J.2.16.6 may be supplemented by C.6.26.10 (AD 531) in which Justinian decides that, where a substitute is appointed to both of two young children, the substitute only comes in if both die before puberty; Justinian is here settling a controversy over interpretation. Ulpian (D.28.6.8.1) makes quite explicit what is implied in paragraph 7, that it is the father's testamentary heir who is substituted if the father says, 'let my heir be the substitute' – meaning the father's testamentary heir in person, not someone who acquires through him, such as the head of his household.

(e) Invalidation of wills
(J.2.17 - 18)

Supervening invalidity (J.2.17). The fact that a will had been properly made and had duly appointed an heir did not mean that it would necessarily take effect as written. Title 2.17 deals with various such cases. A terminological point, made more explicitly in paragraph 5, is that there were overlapping terms for a will which failed to take effect; it might be described as nullified (*ruptum*: literally, 'broken') or frustrated (*irritum factum*: literally, 'made useless or void'), but a will never properly made could also be described as *irritum*. Paragraph 5 suggests clarification by distinguishing explicitly a will not properly made as *non iure factum* (literally, 'not lawfully made') and referring *ruptum* and *irritum* to other cases, *irritum* to cases where the will loses effect through the testator's change of status, and *ruptum* to cases where the will loses effect by some other act or event (see paragraphs 4 and 6(5)). However, it should be noted that Papinian (D.28.3.1) and the rubric to D.28.3 seem to envisage a different scheme: *non iure factum* for a will never properly made, *ruptum* for a will made ineffective by a subsequent act or event, and *irritum* for a will which fails to take effect because the heir does not take up the inheritance (along with a further case, a will of no effect because a son has been passed over). The *Institutes*' distinction between a nullified and a frustrated will could be explained as one between cases where the testator has nullified his own will by subsequent acts, and cases where the will loses effect by operation of law as a result of the change of status which may or may not be the result of some act on his part.

As stated in J.2.17.1, the reason why a will was avoided by adrogation or adoption of a son (full adoption in Justinian's law: J.1.11.2) was that it then became necessary to institute him as heir or disinherit him. The provisions of the will needed to be reconsidered and it was not possible to provide in advance for this case, as it became possible to provide for the subsequent birth of an immediate heir, which would also nullify the will (J.2.13.1 - 3; G.2.130 ff). Papinian (D.28.2.23) indicates this, although in the circumstances (an immediate heir disinherited, emancipated, and brought into the family again by adrogation) the rule is not applied and the disherison stands. As adrogation or adoption was a deliberate act, it was reasonable to require specific reconsideration of the will in the interests of the adopted son. In the time of Gaius, as we learn from G.2.139 -143, there were other cases in which a will could be nullified by bringing a person into *potestas*, such as taking a wife into marital subordination or proving that the marriage out of which the child was produced was entered into in the erroneous belief that the mother was a citizen. In the latter case the law was changed by Hadrian, but only where the error was proved after the father's death, when it was too late to alter the will (G.2.143).

The making of a later will nullified a former one even if the later one failed to take effect, as stated in J.2.17.2, 3. The argument is that the testator has departed from his previous intention and his later will supersedes the former one, whether or not it turns out to be effective, which cannot be known in advance. To the rule that the later will must be properly made there were exceptions. Two are mentioned by Ulpian (D.28.3.2): where the later will is a military one, and where the heir under the later will has a claim on intestacy. Also relevant is C.6.23.21.3 (AD 439), which allows an imperfect will to operate where the heirs are the children of the deceased. Paragraphs 7(6) and 8(7) stress among other things that the general rule applies even where the emperor is named as heir in an imperfect later will. Paragraph 8(7) contains the famous statement (also found Ulpian, D.1.3.31), which became important in later political theory, that the emperor stands above the law, with the significant addition that he lives by it.[20] The reference to 'nomination as heir by mere word of mouth' presumably refers to one without the witnesses needed to make an oral will.

The provision that a will relating only to certain things supersedes an earlier will reflects the rule that appointment of an heir in respect of certain things is treated as appointment to the whole estate, despite the apparent conflict with the testator's stated intention. The heir had to be successor to the whole estate because of the general rule that no one could die partly testate (see section I(d) above). The rule that the former will is totally superseded did not apply if the later will made clear that the heir appointed in it was under trust to pass on the estate to the other heir, less the specific things mentioned and the quarter which, on the analogy of the Falcidian Act, he was entitled to retain (see J.2.22 and J.2.23.5). This is made somewhat clearer in Marcian (D.36.1.30(29)), from which J.2.17.3 is taken. In the *Digest* the passage from Marcian appears in the title on the Trebellian Resolution on trusts of estates (see J.2.23.4 ff).

In classical law lapse of time did not of itself affect a will, except a military will which remained valid for only one year after discharge (J.2.11.3). However, a pronouncement of the emperors Honorius (AD 393-423) and Theodosius II (AD 408-50) of AD 418 (*Theodosian Code* 4.4.6) provided that a will should cease to have effect after ten years, a sensible enough rule in some ways. Justinian alludes to this pronouncement in C.6.23.27 (AD 530) in laying down that the passage of ten years has no effect if the testator has not changed his mind, but that a later will supersedes an earlier one. If the testator has stated that he does not wish his will to stand, or indicated his change of mind in other ways, and this is proved by not less that three witnesses or by a document in the records, and ten years have passed, the will falls by the change of mind combined with the passage of time.

The rule that a will was frustrated by the testator's change of status follows from the fact that normally, as a result of the change, the testator would no longer have an estate or be competent to deal with it by will, for example, if he were adrogated or enslaved. The provision in J.2.17.6(5) allows for the case where the testator is at the time of death capable of having a will and there is a will extant which was originally validly made. In classical law such a will could in effect be saved by granting estate-possession in accordance with it. The praetor overruled the state law where the defect was technical and the testator, having recovered capacity, apparently wished his will to stand, in that he had not made another one. However, the estate-possession was provisional only if there was an heir with a statutory right of succession (G.2.149). Where a person died in captivity, a will made before capture was saved by the Cornelian Act (J.2.12.5). A will of the military fund was not affected by change of status by adrogation or emancipation (J.2.11.5).

Irresponsible wills (J.2.18). A will otherwise valid might also be frustrated in whole or in part by a complaint that it was irresponsible, in the sense that it failed to make adequate provision for members of the testator's family who were regarded as having a claim to such provision. The complaint was a means of obtaining a share of the estate, whereas the rules on disinheriting descendants might simply mean that notice was taken of the descendants in question.

We are ill-informed on the early history of this complaint. Gaius does not mention it, although the institution or something resembling it did exist, as we know from Pliny's *Letters*.[21] The *Digest* title on it (D.5.2) comes immediately before the title on claims for an inheritance, the special remedy given to the heir (D.5.3), and it may be that the complaint was raised by claiming the inheritance on the basis that the will was not effective to exclude the complainant. The absence both of legal sources and of non-legal sources before Pliny leaves considerable scope for speculation. From Pliny it seems that the complaint was brought in the centumviral court, where the old form of procedure by action in the law was still used (see G.4.31). The orators, such as Pliny, who pleaded the case, appear to have argued that there was an element of insanity in the provisions of the will (see Marcian, D.5.2.2, which is the source of J.2.18 pr, from the words 'The form of this... '). The object was to show that in this respect the will could not stand as written, because it failed to make proper provision for the complainant.

In principle the effect of a successful complaint was to set the will aside, and so the complainant had to have a claim to a share of the estate on intestacy in order to benefit. The circle of those who could claim was limited to descendants, ascendants, and brothers and sisters. Only consanguineous brothers and sisters could claim, not uterine (C.3.28.27 of AD 319), and then only against disreputable heirs (described as those who were *infamis* or base or marked as disreputable, or freedmen who were not instituted because of their merit). A slave compulsory heir could not be challenged. More remote relatives who are excluded are mentioned in C.3.28.21 (AD 294). A complaint might be only partially successful if not all the heirs were disreputable. As the complaint involved a reflection on the testator, those who could obtain a share of the estate on the ground that they had not been properly disinherited were required to use the appropriate remedy available by state or praetorian law instead of the complaint.

As J.2.18.3 indicates, the complaint was excluded if an adequate provision was made, the amount being settled at one-quarter of what the complainant would receive on intestacy (see J.2.18.6, 7). This is described as the statutory share (*legitima portio*) and gives its name to the 'legitim' claimed by descendants in some modern systems. If less than this share was given the complaint was not excluded in classical law, but it was usual to provide expressly that the amount given should be made up to the appropriate amount by the appointed heir(s) so as to exclude the complaint. Justinian provided by C.3.28.30.1 (AD 528), 31 (AD 528), and 32 (AD 529), that only those who received nothing could bring the complaint; others could sue only to bring their share up to the appropriate amount

(C.3.28.30 pr), unless they had been given less because they deserved less. The complaint was excluded altogether if the complainant had been justifiably excluded. The shares which had to be given were altered by Justinian in Nov.18.1 (AD 536) in the case of children. Children up to four in number had to share one-quarter of the estate; if more than four in number they shared one-half. In assessing the share to be given, account was taken of all *mortis causa* provisions and of some *inter vivos* gifts (J.2.18.6). Examples are gifts before marriage (C.3.28.29 of AD 479) and gifts specifically intended to be counted towards the share and acknowledged as such (C.3.28.35.2 of AD 531).

As the complaint was that the will was not properly made, acceptance of benefit under the will excluded a complaint and, conversely, making a complaint excluded a person from taking benefit under the will. J.2.18.4 - 5 shows that there was such exclusion only if action was taken personally. The complaint had normally to be made within five years, counted from acceptance of the estate by the heir under Justinian, who also imposed time limits on acceptance to prevent delay in acceptance (Ulpian, D.5.2.8.17; C.3.28.36.2 of AD 531).

The above description of the complaint refers to it before further amendment of the law by Justinian in Nov.115 of AD 542. The precise effect of this pronouncement is disputed, but on one interpretation it fused the law on disherison with the law on the complaint, and provided a unified system of challenge of irresponsible or improperly drafted wills by ascendants and descendants, leaving brothers and sisters under the previous regime.

Before Nov.115 the onus was on the complainant to show that he or she was unjustly excluded (cf. C.3.28.28 of AD 321), but there were no specific grounds of exclusion. According to the *Novel*, children had not merely to be given an appropriate share of the estate but had to be instituted as heirs for at least part of that share, with other relevant benefits counting towards what had to be given in total. If they were disinherited, the parent had to name one or more of fourteen specified grounds indicating ingratitude, such as using violence towards him or her or showing contempt, and the onus was put on the testamentary heir to show that at least one of the grounds stated could be justified (Nov.115.3). Where a child excluded a parent it had to be on one or more of eight specified grounds, such as that the parent had attempted the life of the child, and again the onus was on the heir to establish that a ground existed (Nov.115.4). So far as parents and children are concerned, it does seem that the intention was to combine the provisions on instituting or disinheriting children and those on the complaint, but there are no repealing provisions which might have clarified the intention.

The provisions on irresponsible wills could be evaded or avoided by disposing of the estate *inter vivos* by gift or grants of dowry to persons who were not potential complainants. The practice of doing so led to the development of complaints of irresponsible gifts and dowries, dealt with in

C.3.29 and C.3.30 respectively.[22] The last pronouncement in C.3.29, C.3.29.9 by the emperor Constantius II (AD 337-61) in AD 361, assimilates the complaint of excessive gifts to the complaint of an irresponsible will and the sole pronouncement in C.3.30, again from Constantius II, in AD 358, also speaks of a general assimilation.

(f) Types of heir
(J.2.19)

The treatment of types of heir in this title finds its parallel in G.2.152 - 174 with variations resulting from changes in the law. For example, Gaius says (G.2.153) that a slave compulsory heir must be freed as well as appointed, and he records a difference of opinion on whether the slave should be exposed to ignominy; the wife in marital subordination is one of the immediate and compulsory heirs (G.2.159 - 160), treated as a daughter in intestate succession (G.2.159; 3.3 as restored); and G.2.164 - 173 incorporates discussion of the formal acceptance of the estate called *cretio* and the effect of *cretio* clauses in wills. These were designed to clarify whether or not the inheritance was accepted and to impose time limits on acceptance in the case of outside heirs. Regardless of a *cretio* clause, the praetor might impose a time limit for the sake of creditors (G.2.167) or cut down the time limit set by the testator (G.2.170). *Cretio* was abolished in AD 407 (C.6.30.17; *Theodosian Code* 8.18.8.1). Although the three types of heirs are discussed in connection with testate succession, immediate and outside heirs are equally relevant in intestate succession, as noted in J.2.19.2. It is worth stressing the point made in J.2.19.3 that 'outside' heirs are not necessarily outside the circle of relatives of the deceased; they are outside the sphere of *potestas*. In connection with the various types of heir it is also worth noting that a person who was entitled to succeed under a will and was also entitled to succeed on intestacy was prevented by the praetor from taking up the estate as intestate heir so as to defeat the claims of other persons intended by the testator to benefit under the will, such as legatees or slaves granted freedom (see D.29.4).

One slave could always be freed to provide a compulsory heir, although this removes the value of the slave from the inheritance to the prejudice of creditors. This was expressly provided for in the Aelian-Sentian Act (J.1.6.1). The separation of his acquisitions after the death, referred to at the end of J.2.19.1, was a concession which had to be applied for, to the praetor in classical law (Ulpian, D.42.6.1.18).

The idea that immediate heirs are already in a sense owners also appears in Paul, D.28.2.11. It may reflect an early notion of family property but it has no reality in the law as we know it from the *Twelve Tables* onwards. The right to stand off protected the acquisitions of immediate heirs after the death in a way comparable to the separation granted to slave compulsory heirs. It meant that, even in the case of immediate heirs, the question could arise whether the inheritance was accepted, because the right to stand off could not be claimed if the heir meddled with the estate (cf. Pliny, *Letters* II.4). The praetor gave time for deliberation, and 'meddling' meant doing some act which implied acknowledgment of the burden of the inheritance. Acts of piety, such as seeing to the funeral, did not count. The heir could change his mind at any time if the estate had not yet been sold, but by C.6.31.6 (AD 532) Justinian allowed a maximum of three years (except for minors).

Outside heirs were regarded, in general, as being protected from the risk of accepting an insolvent inheritance by having the opportunity to consider whether to accept or not. The praetor in classical law could impose a time limit in the interests of creditors (G.2.167). In AD 529 Justinian gave them a year to deliberate and allowed the right of acceptance (exercisable in the rest of the year) to pass to their successors if they had not renounced (C.6.30.19). Then in AD 531 by C.6.30.22 he made the radical change referred to in J.2.19.6, introducing the benefit of inventory. An heir who was doubtful of the solvency of an inheritance but did not reject it (within three months of knowing of his right) could make an inventory of the estate, normally to be begun within thirty days of his knowing of his right and completed within a further sixty days; if he did so he was given the benefit of the Falcidian Act (J.2.22) and his liability to creditors was limited to the estate inventoried. Only an heir who failed to make an inventory but accepted the inheritance was now fully liable.

While an outside heir was making up his mind whether or not to accept, the inheritance was said to be *iacens* (literally, 'lying (dormant)'). It had no owner for the time being; taking something from it was not theft and originally it could be taken by anyone and acquired by usucapion in a year (G.2.54 - 58).[23] However the idea developed that the inheritance itself was a sort of person or represented a person, so that, for example, acquisitions could be made for it through slaves belonging to the inheritance, and damage done to property in the inheritance could be sued for once the heir accepted.[24] The inheritance is generally referred to as representing the person of the deceased (J.2.14.2; Ulpian, D.41.1.34) but it was held by some that it represented the person of the heir who accepted. In any case it appears that so far as the inheritance could represent a person, the effect was retrospective, in that the heir acquired what had been acquired for the inheritance (Paul, D.50.17.138 pr). There is no totally consistent doctrine, and that there were limitations is indicated by J.3.17 pr. It is going too far to suggest that the inheritance was treated as a legal person, although the germ of such a doctrine is there.

With the abolition of formal acceptance by *cretio*, informal acceptance, which even in the time of Gaius could take the form of acting as heir (G.2.163, 167), was of more importance and Justinian's *Institutes* have more discussion of what counts as acting as heir than has Gaius, who gives no examples. The reference in J.2.19.7 to 'merely ... showing an intention to accept' presumably refers to informal acceptance for Justinian, but the phrase also appears in G.2.167, from which it is expunged by some scholars as a gloss. The intention would have to be proved; the reference can scarcely be to an unexpressed intention and, if so, it is not clear that Gaius did not say what is attributed to him.

(g) Legacies
(J.2.20 - 22)

Three titles deal with legacies but J.2.20, on their constitution and interpretation, is by far the longest of the three. Its internal structure is not altogether satisfactory, partly because it was built up from existing material which was adapted to its new situation and which reflects questions which were no longer as relevant or important as in the earlier law. For example, the different types of legacy discussed by Gaius in a long passage (G.2.192 - 223) at the start of his treatment were deprived of significance by imperial pronouncements, including the very sweeping one of Justinian in C.6.43.1 (AD 531) referred to in J.2.20.2.

The definition of a legacy in J.2.20.1 is too general to be helpful. In D.31.36 (Modestinus) a legacy is said to be a gift left by will, but even in classical law a legacy could be left by a codicil confirmed by will (G.2.270a). With Justinian's equation of legacies and trusts, effected by C.6.43.2 (AD 531) and referred to in J.2.20.3, legacies could be left even by unconfirmed codicils, as could trusts before (G.2.270a; J.2.25.1), and this may explain the absence of reference to a will in J.2.20.1. The forms of legacy referred to in J.2.20.2 are discussed in some detail in G.2.192 - 223, but even by Gaius' time the importance of the various forms was substantially diminished by the Neronian Resolution (between AD 54 and 68) which allowed legacies improperly formulated to be treated as if left in the most favourable way, that is, by obligatory legacy (G.2.197 - 198, 218). A properly formulated and competently made proprietary legacy gave ownership to the legatee directly (G.2.194) whereas the other main form, the obligatory legacy, gave the legatee a personal action against the heir. Justinian's pronouncement in C.6.43.1 (AD 531) allows a legatee both a profusion and a confusion of remedies: an action to claim ownership (in appropriate cases), a personal action against the heir, and an action to claim security over the heir's estate for payment of the legacy. Traces of the older distinction linger in the *Digest*.

The equation of legacies and trusts is referred to in D.30.1 in a text attributed to Ulpian, but this text has long been recognised as interpolated. However, the institutions of legacy and trust were being brought closer together even by the time of Gaius, who notes both the existing differences (G.2.268 - 283) and the differences removed, by imperial pronouncements or otherwise (G.2.284 - 288). In spite of the equation, Justinian retains separate titles on legacies and trusts in the *Institutes* and in the *Digest* and *Codex*. The separate treatment of trusts of estates (J.2.23; D.36.1; C.6.49) is easily understandable and readily justifiable. The separate treatment of legacies and trusts of individual things really reflects the way in which the *Corpus Iuris* was compiled, although Justinian does offer some excuse in J.2.20.3 for the treatment in the *Institutes*.

After the introductory three paragraphs the topics discussed in title 2.20 can be distinguished, following roughly the order in which they appear, as

(1) the possible object of a legacy, dealt with somewhat incoherently, in paragraphs 4 - 7, 9 - 10, 13 - 15, 17 - 23;
(2) implied ademption of legacies: paragraphs 6, 9, 12 (see also the end of paragraph 10);
(3) joint or several legacies: paragraph 8;
(4) error by the testator: paragraphs 11, 29 - 31;
(5) risk of destruction of the object of a legacy including manumission of a slave: paragraphs 16 - 18;
(6) questions of interpretation of the extent of a legacy (overlapping with discussion of the object of a legacy): paragraphs 17 - 23;
(7) capacity to take a legacy: paragraphs 24 - 28, 32 - 33;
(8) abolition of the earlier rules on the placing of legacies in a will: paragraph 34; the invalidity of legacies after the death of heir or legatee: paragraph 35; and the invalidity of legacies made, adeemed, or transferred as a penalty on the heir: paragraph 36.

Several of the topics are expanded considerably in the *Digest* (D.30 - 36) and *Codex* (C.6.37 - 54).

The last two subjects require very little comment and can be dealt with immediately. The sections in question deal mainly with the removal of restrictions found in classical law. Some of these had borne harder on legacies than on trusts, but there was a process of assimilation evident in G.2.284 ff. Where unidentifiable persons are concerned there does not appear to have been the same difficulty over leaving legacies to bodies such as towns as there was over their appointment as heirs (see section (d) above). Legacies and trusts in favour of towns were common. It was no doubt easier to allow them to accept a specific benefit than to allow them to act as heirs, as the membership would be fixed at the time of vesting of the bequest.

The rather mixed-up order of treatment of the possible objects of a legacy reflects the fact that in classical law what could be left by legacy depended in part on the type of legacy chosen. For example, a thing belonging to someone else could not be left by proprietary legacy (at least to the effect of transferring ownership in it; the legacy as such might be saved by the Neronian Resolution). The treatment in the *Institutes* reflects that of Gaius, who begins by dealing with the different forms of legacy and discusses as specific to the different forms of legacy various points (including joint legacies) which Justinian treats as general ones. Despite the range of remedies given to the legatee by Justinian, which includes a right to claim ownership, a legacy need not be of a corporeal thing, as is stated specifically in paragraph 21 and implied elsewhere, such as in J.2.20.13 and 14.

Implied revocation of legacies is a more intricate topic than the express revocation dealt with in J.2.21. It involves difficult questions of interpretation, whether the original object of the legacy belonged to someone else, as in J.2.20.6, 9, and 10, or to the testator himself, as in J.2.20.12. In the case of *mortis causa* acts, questions of interpretation are the more difficult to resolve in that, even if the law permits evidence of the party's actual intention to supplement the wording of the document giving rise to the problem of interpretation, that evidence may no longer be available. Parole evidence clearly is excluded, so that any evidence must be indirect. Whether evidence of the deceased's intention other than the wording of the actual document to be construed will be admitted is another question, and one on which Roman law was more liberal (or looser) than modern Scots and English law. That there was a dispute over implied ademption is indicated particularly by J.2.20.12.

The rule on joint or several legacies stated in J.2.20.8 is that stated by Gaius for joint or several proprietary legacies (G.2.199). In a joint obligatory legacy, if one legatee failed, his share reverted to the heir (G.2.205) (subject to the effect of the Papian Act, which treated that share as caduciary and gave it to the other joint legatee or legatees having children; see G.2.206 - 208). The treatment of joint legacies may indicate a concept of joint ownership as ownership of the whole limited by the ownership of the other joint holders, or may simply be a matter of interpretation of the grant.

The forms of error discussed in title 2.20 are error about the ownership of a thing bequeathed (paragraph 11), error about the name of a legatee (or heir) (paragraph 29), and error in the description of the object of the legacy (paragraph 30) or in the stated reason for bequeathing it (paragraph 31). Paragraph 11 should be considered along with paragraph 4, giving the general rule on legacy of another person's property. The statement at the end of paragraph 31, that the legacy fails if it is expressly made conditional on the truth of the reason stated, shows that questions of error are always, in part at least, questions of interpretation. The wrongly stated reason could have been interpreted as a condition; instead, the fact that a legacy was given is treated as more significant than the fact that a false reason was given (and, of course, the testator's actual intention cannot be explored by asking him, as would be possible in an *inter vivos* transaction). The same approach is taken to a false description (paragraph 30). The law does not leave the issue to be resolved by reference entirely to the subjective intentions of the party in error.

It may be noted that the question whether an heir is liable to the legatee if he is unable to transfer the object of a legacy to him depends not on fault on his part, but on whether the impossibility came about through his act (paragraph 16). The value of a manumitted slave is not claimable because the slave is now free and has no commercial value. The risk of reduction of the value of the estate between death and payment of legacies fell on the heir (J.2.22.2).

Title 2.20 illustrates various questions of interpretation, particularly in paragraphs 17 - 23, and it is clear from these paragraphs, and from the more extensive discussion of questions of interpretation in *Digest* texts and in the *Codex*, that interpretation of wills caused as much difficulty to Roman as it does to modern lawyers. It is doubtful if there is any consistent doctrine of interpretation in the texts, and it is simplistic to suppose that there was a progressively greater stress on the importance of the testator's intentions after the famous *causa Curiana* of the late Republic. In this case the orator L. Licinius Crassus successfully argued for upholding a faulty substitution for a child on the basis of an interpretation based on the testator's intention, against a strict interpretation for which the jurist Q. Mucius Scaevola argued. His success is sometimes seen as marking the victory of the freer style of interpretation of the orators over the narrower and more pedantic interpretation of the republican jurists, but this reads too much into the case.[25] There are certainly differences of opinion recorded among classical jurists which indicate that a narrow interpretation can be found relatively late in the classical period; it may also be doubted whether individual jurists always followed the same line.[26] The interpretation of a particular provision is in any case always liable to reflect the particular circumstances in which the question arises. At the same time it is clear that there is stress on the intentions of the testator in the *Corpus Iuris*, some of it attributable to the compilers (cf. J.2.20.3, 34), and there was some tendency to give an interpretation which would uphold the actual or supposed intentions of testators and to favour freedom in interpreting questions of manumission (J.2.20.34).

A legacy or trust which was in principle valid might be forfeited for unworthiness (see section II(a) above). If legacies exceeded the estate as at the date of death they were cut down proportionally; the Falcidian Act then applied to ensure that the heir had some share of what was left (J.2.22.2, 3).

Revocation and transfer of legacies (J.2.21). This brief title deals with express revocation of legacies and their transfer from one beneficiary to another. It appears that at one time it was important to use the correct wording to achieve the result desired, just as it was important to use the correct wording to make a legacy. Informal revocation was always competent for trusts (Ulpian, D.34.4.3.11, 4), and from this text it appears that there could be implied revocation of a trust, say by serious breach of friendly relations, where a legacy would stand by state law but an action to enforce it would be met by a plea in defence.

The Falcidian Act (J.2.22). Gaius (G.2.224 - 227) gives somewhat fuller details of the Furian and Voconian Acts. The defect of both was that, while imposing limitations on the amount of individual legacies, they did not limit the number of legacies which could be left. The heir, therefore, might still be deprived of any substantial benefit. Gaius gives no details of the operation of the Falcidian Act in his *Institutes*, but J.2.22.2 and 3 may be compared with D.35.2.73 pr and 73.5, both from Gaius. The act did not apply to military wills (C.6.50.7 of AD 226) but did apply to trusts (G.2.254; J.2.23.5). On overclaim by a legatee, see J.4.6.33.

It may be deduced from Ulpian, D.35.2.64, that a testator could allow particular legatees to get their full legacies at the expense of others so long as the heir got his quarter, but he could not prohibit the heir from taking his quarter. By Nov.1 (AD 535) Justinian made new provision. On the one hand, he laid down that if the heir thought that legacies etc. exceeded three-quarters of the estate he must call together the beneficiaries or their representatives and make up an inventory of the estate, and he could then retain his quarter; if he failed to follow this procedure he must pay in full, even beyond the value of the estate. On the other hand, Justinian allowed the testator to declare that the heir should not be entitled to retain his Falcidian quarter. An heir who accepted in spite of this would in effect become merely an executor (see section I(c) above), and Nov.1.1 provides for the carrying out of the provisions of wills by other beneficiaries if the heir refuses to do so. It also became possible to ask an heir to pass on the whole estate to a successor by a trust of the estate (see section (h) below).

(h) Trusts and codicils (J.2.23 - 25)

To his account of trusts of estates in J.2.23 Justinian prefaces some remarks on the history of trusts, which can be supplemented by what is said in J.2.25 on the history of codicils. Gaius, whose account of trusts of estates in G.2.248 - 259 supplies much of the context of J.2.23.2 - 6, says nothing of the history of trusts beyond the remark in G.2.285 that trusts more or less began with trusts in favour of foreigners. From Cicero, *de finibus* 2.17.55 and 2.18.58 and II *in Verrem* 1.47.123 ff, we know that trusts were used by testators before they became legally enforceable, but there is some variation between the two accounts given by Justinian of how they did become legally enforceable. In J.2.25 pr he is more precise on the role of Lentulus who may have set a fashion, if we may judge from J.2.23.1, leading Augustus to discuss the question of enforceability with the jurists. It became a common practice to appoint emperors as legatees and sometimes a bequest was made in the hope that the emperor would validate a defective will in order to benefit; the temptation was apparently resisted.[27] The special praetor referred to in J.2.23.1 was created under Claudius (although the consuls did retain a jurisdiction) and we know from G.2.278 that the special judicature continued to be used. Trusts fell outside the ordinary courts and their formulary procedure. They were used particularly to do things which could not be done in the ordinary way by will, such as benefiting non-citizens, but the scope for evading the normal rules was narrowed, as we are told by G.284 - 288. Once they became legally enforceable a narrowing of their scope is only to be expected, as they could then hardly be used to effect purposes which were regarded as contrary to the policy of the general law, for example, to evade the Augustan legislation against the childless and unmarried.

Trusts of estates (J.2.23). The observation in paragraph 3 is more fully explained in paragraphs 4 - 6 and by the corresponding paragraphs in Gaius. From G.2.252 we learn that, before the Trebellian Resolution (AD 56) the device used to transfer the estate to the beneficiary was to treat him as a nominal purchaser of the estate, and to use the reciprocal stipulations entered into between an actual purchaser of the estate as a whole and the heir as seller. The efficacy of such a device depended, however, on both parties remaining solvent. Where no more than three-quarters of the estate was to be handed over the Trebellian Resolution made this device unnecessary. The reciprocal agreements allowing actions and pleas in defence of claims in respect of the estate were replaced by policy actions which G.2.253 says appeared in the praetor's edict. The reciprocal undertakings, therefore, fell out of use in this case. However, when the Pegasian Resolution (of ca. AD 73) allowed the heir to retain one-quarter, the clumsy device of reciprocal undertakings in respect of three-quarters and one-quarter of the estate was used, on the model of undertakings between an heir and a cut-in legatee, who was left a share of the inheritance (J.2.23.6; G.2.254). Moreover, where the heir voluntarily handed over the whole estate, the undertakings as between seller and purchaser were used instead of the neater policy actions of the Trebellian Resolution (G.2.257). Only if the heir was compelled to enter in terms of the Pegasian Resolution (as he could be if he refused, claiming that the solvency of the estate was doubtful; see G.2.258), was there automatic transfer with policy actions on the Trebellian model.[28] Clearly, Justinian's assimilation of the provisions (J.2.23.7, not preserved in the *Codex*) was a sensible improvement. An important distinction is made in J.2.23.9 between deducting an individual thing or things or a sum of money for the benefit of the heir, and deducting the quarter to which he was entitled. The effect of Nov.1 was to allow the testator to prohibit the retention, as in the case of legacies and individual trusts.

Paragraph 11 of title 23 refers to the possibility of the trust beneficiary in turn being requested that the estate or part of it be passed on. The removal by Justinian of the earlier prohibition on naming beneficiaries who were not definitely ascertained is generally thought to have made it possible for the testator to name successive beneficiaries in a trust disposition without limit. If he also prohibited retention of a quarter, the whole estate could then be passed on in perpetuity undiminished (except by the accidents of time) as contemplated by the fee tail of English law or the entail of Scots law. Whether he was so radical is doubted, but in any case Nov.159.2 (AD 555) limited such successive substitutions to four.[29]

From J.2.23.12 we learn incidentally that a trust normally required writing or five witnesses. Justinian's provision of AD 531 allowed proof by remission to the oath of the alleged trustee. This and other provisions referred to in paragraphs 10 and 11 applied not only to trusts of estates but to trusts of individual things. Similar points are made by Gaius (G.2.270 and 271), but better placed among the general differences between trusts and legacies.

Trusts of single things (J.2.24). The title on trusts of single things requires little comment. Most of it is taken from G.2.260 - 267, but J.2.24.3 corresponds to G.2.249 and there are differences. G.2.262, for example, which corresponds to the end of J.2.24.1, refers to a dispute over the effect of a refusal to sell by the owner of a third party's property left by trust. Some jurists held that the trust failed, although an obligatory legacy of the property of a third party was still enforceable and the heir had to pay the value. Again, G.2.265 says that a grant of trust freedom to a slave of a third party fails if the owner of the slave refuses to sell him. The rule in J.2.24.2 appears in C.7.4.6 (attributed to Alexander ca. AD 222). A person who received a benefit from the testator could be compelled to free his slave if asked to do so by a trust. Gaius (G.2.267) also gives examples of wording of a direct grant of freedom and refers not merely to ownership but specifically to quiritary ownership, a refinement unnecessary for Justinian's law, in which there was only one form of ownership (C.7.25.1 of AD 530-1).

The title refers only to trusts of corporeal things (and of grants of freedom). There could also be trusts of incorporeal things and, in particular, of a usufruct.[30] Trusts of usufruct were of considerable social importance as a means of providing for persons who could not be benefited directly.

Codicils (J.2.25). The term 'codicil' (literally a little *codex* or book) is rather misleading for Roman law, in so far as in English, the term 'codicil' carries the implication of an appendix or supplement to a will. In Roman law a codicil was simply an informal testamentary document (J.2.25.3) which had no necessary connection with a will but was commonly used to create a trust (cf. J.2.25.1). It could always be in Greek (G.2.281), while legacies had to be in Latin until AD 439 (C.6.23.21.6). However, a distinction was drawn between a codicil confirmed by a will and a codicil not so confirmed (Paul, D.29.7.8 pr), a distinction more important for classical law than for the law of Justinian. An unconfirmed codicil in classical law could only create a trust. A confirmed codicil could do anything which a will could do, except directly create an heir, provide a substitute heir, change the terms of the heir's appointment or disinherit someone (G.2.273; J.2.25.2), or appoint a guardian (G.2.289). Even in Justinian's law confirmation could be important, because an independent testamentary act normally required five witnesses (except for a will which normally required seven) (C.6.36.8.3 of AD 424). Paragraph 3 refers to formalities of constitution. The authenticity of an alleged codicil had to be proved: see Justinian's pronouncement of AD 530 in C.6.23.28.6 requiring five witnesses to that effect.

The existence of codicils as informal testamentary writings created the same sort of problems as do informal testamentary writings today, because there could be a question whether they stood along with a will or not, if they were not expressly confirmed. Papinian's view recorded in J.2.25.1 would produce a clear answer. The written reply of the emperors Severus and Antoninus, which is not preserved in the *Codex*, does not, although it gives more weight to the testator's intention. Other texts show that a testator might himself lay down the conditions on which a codicil should be given effect, for example, that it should be subscribed or sealed (Marcian, D.29.7.6). Marcian (D.29.7.6.2), who says that codicils not complying with such requirements are still valid because later provisions derogate from earlier ones, seems to imply that a confirmation clause might be read as applying only to codicils already made before the will.

A will which failed to take effect as a will might take effect as a trust, but only if this was expressly provided (C.6.36.8.1a of AD 424; Ulpian, D.29.7.1).

III. Intestate Succession on Death
(J.3.1 - 9)

The treatment of intestate succession on death is considerably briefer than the treatment of testate succession, both in the *Institutes* and in the rest of the *Corpus Iuris*, partly because it was less complex and partly because some matters (such as types of heir) which are discussed in relation to testate succession were common to both branches of the law. It does not necessarily reflect any preference or particular concern for testate succession. The amount of change in the details of the law of intestate succession over the history of Roman law might, indeed, suggest the contrary. Intestate succession came into play when there was either no will or no effective will. It might not be immediately clear that there would be intestacy (see G.3.13; J.3.1.7; 3.2.5), and the successors were determined as at the date when it was settled that the inheritance was to be distributed on intestacy (J.3.1.8; 3.2.6), although the successors must be in life or at least conceived by the date of death (J.3.1.8). As already noted (section I(d) above), in general there could not be a mixture of testate and intestate succession as there can in modern law, although there were exceptions.

4. Succession

Four main phases of development may be distinguished: the early law resting on the *Twelve Tables*; the changes made by the praetor; imperial innovations, including the detailed changes made by Justinian in the *Institutes*, *Digest*, and *Codex*; and finally the recasting of the law by Justinian in the *Novels*. A general trend is the increasing recognition of blood relationship, as opposed to the narrower agnatic relationship which rules the early law. This leads to the substitution of the principle of blood relationship for agnatic relationship in Justinian's final scheme. For all the changes in detail, the basic structure and order of succession – immediate heirs; agnates; cognates – stayed until the introduction of that final scheme (cf. J.3.6.11 - 12). What has been said applies primarily to succession to free men and women. Freedmen were different in a number of respects because of the claims of their patrons and, in classical law, because there were different kinds of freedmen. Succession to them is dealt with separately.

(a) Succession to free men and women
(J.3.1 - 6)

Inheritance on intestacy (J.3.1). The general heading of this title is rather misleading. It deals, not with the generalities of intestate succession, but with the persons who had first claim to the inheritance, of whom there were three groups: (1) by state law, the immediate heirs, who were given first place by the *Twelve Tables* (J.3.1.1 - 8); (2) by praetorian law, emancipated children who were not in another family (J.3.1.9 - 11); (3) by imperial pronouncements, direct descendants through females (J.3.1.15 - 16).

Compared with the list of immediate heirs in G.3.1 - 8, there are three omissions: (1) wives in marital subordination (G.3.3); (2) children brought into *potestas* under the Aelian-Sentian Act (G.3.5) (as extended by senate resolution, see G.1.29 ff), or in other ways not specifically mentioned by Gaius at this point, such as by proof of error (G.1.66 ff) or by imperial grant (G.1.92); and (3) a son not completely freed from *potestas* in the process of emancipation or adoption because not mancipated for the three times which were necessary to release him (G.3.6; cf. G.1.132). Additions are children legitimated as mentioned in J.3.1.2a. The variations of course reflect changes in the law and in the procedure for emancipation and adoption (see Chapter 2, sections II and IV above). Nov.74 (AD 538) introduced legitimation by imperial pronouncement, to add to the cases in J.3.1.2a.

In proceeding from the rights of immediate heirs to the rights of emancipated children, given by praetorian law, Justinian departs from Gaius' order of treatment. Gaius discusses the praetorian rules only after full exposition of the state law (G.3.18 ff). This gave the next claim after immediate heirs to the nearest agnate, and finally gave the inheritance to the gentiles, members of the *gens* or clan, whose rights were already long obsolete in Gaius' time (G.3.17). The context in Gaius is the improvements effected by the praetor in the narrow scheme of the *Twelve Tables* (G.3.18 - 25). Justinian's treatment of the position of emancipated and adopted children is fuller, and he explains his alteration of the law of adoption (J.3.1.14; cf. J.1.11.2) as a measure to protect adopted children. The effect was to give to the adopted child only rights on intestacy in his adoptive father's estate unless there was full adoption. The relevant pronouncement, C.8.47.10 (AD 530), by subparagraph 3, abolished the special rule of the Afinian Resolution referred to in J.3.1.14, which apparently guaranteed one-quarter of the estate where one of three brothers was adopted. Nothing is known of this resolution beyond what we are told by Justinian and by Theophilus in his paraphrase of the *Institutes* at J.3.1.14.

Emancipated children, particularly in the classical period, had greater opportunities of acquiring property of their own than had those in *potestas*, who might at best have a military fund. Emancipated children, therefore, could not claim on intestacy without offering to bring in a share of the property which they had acquired since emancipation, that is, to collate a due share of their property. By Justinian's law it was necessary to collate only what had come from the head of the family, because by that time members of the family could hold as their own, property which had come from other sources (see Chapter 2, section II above). There was therefore no longer the same unfairness in allowing a claim without contribution of property from sources other than the head of the family. The duty to collate applied only where claims were made both by those who had been under *potestas* and by those who had not (C.6.20.9 of AD 293). There was no question of equalisation of the position of claimants who had all had the opportunity to acquire property independently of the head of the household. Dowry and gifts on account of marriage in particular had to be collated, but in Justinian's law all gifts counted (C.6.20.20 of AD 529). Previously it seems that only gifts intended by parents to be collated had to be brought in.

The imperial innovations referred to in J.3.1.15 and 16, giving rights to descendants through females, are a breach with the agnatic principle of the state law, but they are in line with the general tendency to increase the rights of cognatic relatives. They also foreshadow Justinian's ultimate scheme in the *Novels* (described in section (d) below).

Statutory succession by agnatic relatives (J.3.2). By state law, in the absence of immediate heirs the nearest agnatic relative took the inheritance, and the praetorian law did not displace him or her. However, the class of nearest agnate was quite narrowly defined. Justinian's account of the older law parallels that of Gaius in 3.9 - 16, 23, and 24, but he often paraphrases the text of Gaius. An element of discrimination against females was introduced by denying a claim as agnates to female agnates beyond sisters (G.3.23; J.3.2.3); they could claim only as cognates in the next class under the praetorian scheme. Justinian removed the discrimination against females, reverting he says (J.3.2.3b) to the scheme of the *Twelve Tables*. He improved the position of nephews and nieces claiming through their mother to an uncle's estate by giving them a claim in the class of agnates (J.3.2.4), but his introduction of succession among agnatic claimants (J.3.2.7) pushed back the claims of cognatic relatives, who had been benefited by the narrow state-law definition of the nearest agnate.

The praetorian scheme of intestate succession operated by the grant of estate-possession on intestacy (J.3.9.3 - 8(7)). While agnates came in the second class of claimants (the first being immediate heirs and children included with them as described in the first title), that class was described more generally as those with a statutory claim. It therefore included cases other than agnatic relatives. J.3.2.8 refers to one of these cases, where a head of a household retained a statutory right of succession despite emancipating a descendant and so breaking the agnatic relationship between them. This was done by entering into an arrangement that after the child had been mancipated to another party in order to break the head of the household's *potestas*, that party would not release the child as he could do in the absence of such an arrangement, but would mancipate him or her back to the head of the household to be released. This gave the head of the household a statutory right of succession, which otherwise went to the party manumitting the child in the final part of the procedure for emancipation (cf. G.1.132, which unfortunately is mutilated). In simplifying the procedure of emancipation Justinian gave all parents a right of succession automatically (C.8.48(49).6 of AD 531; J.1.12.6; 3.9.5).

The Tertullian Resolution (J.3.3).[31] Except where a wife was in marital subordination and treated in succession as agnatic sister to her own children (and these cases were increasingly rare), a mother had no right to succeed to her children, except as a cognate under the praetorian scheme of succession (G.3.24 - 25; J.3.3 pr). The Hadrianic Tertullian Resolution is the first permanent provision improving her position. The right given by Claudius, referred to in J.3.3.1, seems to have been a personal privilege. We have no further information from Gaius because the passage in which he appears to have dealt with the Tertullian Resolution in the *Institutes* (after G.3.33a) is lost. (He also wrote a separate commentary on it and on the Orfitian Resolution, from which there are excerpts in D.38.17.8, 9.)

The right of the mother was originally tied to the privilege of children given by the Julian-Papian Act (see Chapter 2, section V(f) above). According to C.8.58(59).1, the privilege of children was given to all mothers by the emperors Honorius and Theodosius in AD 410 but, as Justinian expressly gives a right to mothers with one child (J.3.3.4; C.8.58(59).2 of AD 528), either the earlier grant was not as broad as it seems – perhaps because of interpolation – or the resolution was interpreted as still governing entitlement to rights of succession.

The Tertullian Resolution appears to have made the mother heir to the child and not merely estate-possessor (cf. J.3.4.2) and so she competed with the agnates as a statutory heir (and was also entitled to estate-possession as such). Her position as statutory heir was not affected by change of status, however (J.3.4.2), although change of status destroyed agnatic relationship. As J.3.3.5 indicates, the mother did not always come in before the agnates for the whole estate, but sometimes shared with them. She was always postponed to immediate heirs and children (if any), which meant that she was in an even worse position in relation to sons than daughters, because sons could have immediate heirs while daughters could not, as they had no *potestas*. Justinian favoured her by giving her half the estate against sisters but required her to share proportionately with brothers (C.6.56.7 of AD 528); in Nov.22.47 (AD 536) he provided that she should also share proportionately with sisters. In the *Institutes* he does not deal with competition between a surviving father and mother. In the *Codex* he provides that where both parents and brothers and sisters survive, the parents get a usufruct of two-thirds divided between them; the usufruct of the remaining third and the property in the whole estate go to brothers and sisters. Where the child is still in *potestas*, the father keeps the usufruct which he had when alive (J.2.9.1), while the mother comes in with brothers and sisters as described above. All this is without prejudice to the effect of a mother entering into a second marriage.

From Ulpian (D.38.17.2.23 ff) it appears that the penalty imposed for failing to apply for a guardian as soon as possible (J.3.3.6) was imposed in the resolution itself. The rule that a mother could claim on the estate of illegitimate children also applied in the Orfitian Resolution (J.3.4.3).

The Orfitian Resolution (J.3.4).[31] The Orfitian (or Orphitian) Resolution (AD 178) is referred to in Gaius (D.38.17.9) as a speech of the emperor (*oratio*), implying that the resolution of the senate was a formality. The grant of a right of succession by children to their mothers led, under the emperor Constantine (cf. C.6.60.1 of AD 319), to the recognition of property derived from the mother as a separate estate (referred to as *bona materna*, literally, 'maternal property'). The extension of the right to grandchildren referred to in J.3.4.1 occurred in AD 389 (C.6.55.9; see also C.6.55.11 of AD 426). Before the resolution children had a right as cognates related in the first degree (J.3.6.2), but were excluded by agnates; a mother had no immediate heirs, as she had no *potestas*. The purpose of the Orfitian Resolution was to cut down the claims of agnates in favour of children who were made statutory heirs (J.3.4 pr) with a claim not affected by change of status (J.3.4.2).

The rule given in J.3.4.3 is qualified by Justinian in relation to mothers of the highest rank (*illustres*) by C.6.57.5 (AD 529). He provides that where such a mother has both legitimate and illegitimate offspring (with no known father) the illegitimate children get nothing, because a woman of that rank ought to be chaste.

Where there was a competition between claimants under the two resolutions because a child left both a mother and children, who were both in the first degree of relationship to the deceased as cognates (J.3.6.2), it appears that the children of the deceased child were preferred. In C.6.57.1 (AD 225) a daughter is preferred to the mother and full agnates.

J.3.4.4 appears to be misplaced in the treatment of the Orfitian Resolution. It comes from Marcian's *Institutes*, book 5 (D.38.16.9) and states a general point. It may appear where it does because both resolutions diminish the right of other statutory heirs.

Succession of cognatic relatives (J.3.5). The praetor introduced the right of cognatic relatives to estate-possession. Gaius (G.3.27 - 31) gives further details of cognatic succession in default of the nearest agnate, but he does not mention the restriction on the degrees of relationship of agnatic or of cognatic relatives who could claim, referred to by Justinian (J.3.5.5(4)). Cognatic relationship survived change of status or status-loss and so an agnate who lost his or her statutory claim as a result of status-loss could still claim as a cognate, albeit this claim as cognate was postponed to that of remaining agnatic relatives and (s)he might have to share with a larger number of claimants (G.3.27; J.3.5.1). An immediate heir who had been emancipated might still be able to claim as a child (cf. J.3.5.1 and C.5.30.4 of the emperor Anastasius, AD 498), but was in any case a cognate if no claim was made as a child. A child adopted into another family was still a cognate (G.3.31).

Although Justinian still speaks of the Anastasian Act in the present tense (J.3.5.1), the act is not found in the second edition of the *Codex*, although it is referred to in another pronouncement of Anastasius preserved in C.5.30.4 (AD 498). It was overtaken by Justinian's own provision in C.6.58.15.1 (AD 534), supplementing C.6.58.14 (AD 531), which equated male and female in the succession of agnates (see J.3.2.3). He equated the position of emancipated brothers and sisters with the position of those in *potestas*. Justinian also gave uterine brothers and sisters a claim as statutory heirs (C.6.58.15.2 of AD 534).

Degrees of cognatic relationship (J.3.6). The principle behind the calculation of degrees of relationship explained in this title (a calculation which also applied to agnatic relationship but counting through males: J.3.6.8) is that each step up towards a common ancestor counts as one degree of relationship and so does each step down (J.3.6.7). In contrast, the Canon law computation counts the longer of the lines of descent from the common ancestor as determining the degree of relationship. The diagram referred to in J.3.6.9 is not found in the best manuscripts but is often supplied in medieval and later editions of the *Institutes* as a simple diagram or in more fanciful form, for example, as a tree. Paul (D.38.10.10) gives further details, and the text of J.3.6.1 - 6 is close to that of Gaius, D.38.10.1 and 3 pr.

The reforming constitution referred to in J.3.6.10 is C.6.4.4 (AD 531) on the estates of freedmen, which was promulgated in Greek so that it could be more widely known (J.3.7.3). As noted in the following title, Justinian greatly simplified the law regarding succession to freedmen, in particular by providing that all freedmen should become citizens (see Chapter 2, section I(c) above).

(b) Succession to freedmen and freedwomen (J.3.7 - 8)

Even if freed slaves became citizens, which not all did in classical law (see Chapter 2, section I(c) above), succession to them followed a different pattern from succession to a person born free. On the one hand, although a freed male slave could subsequently have immediate heirs (and children freed from *potestas*) as claimants to his succession, no freed slave could have other agnatic or cognatic relatives. On the other hand, the patron or patroness who manumitted the slave, and children of the patron, had statutory rights of succession on intestacy under the *Twelve Tables*, which were extended by the praetor in the case of patrons and their male descendants (G.3.41, 45 - 46). These included a right to challenge a will which did not give a share to the patron or his male descendants. J.3.7 pr - 2 is taken from G.3.39 - 42, with omission of Gaius' references to the wife or daughter-in-law in marital subordination along with the adoptive child. The improvement in the patron's position (and in the position of a patroness) under the Papian Act was part of the Augustan legislation to encourage a higher birth-rate, especially among the upper and wealthier classes who tended to produce fewer children than the poor.[32]

Gaius in G.3.43 - 76 gives a long account of other cases of succession to freedmen and freedwomen. Freedwomen were in a different position from freedmen because, unless dispensed from guardianship by the Papian Act when they had the privilege of children (G.3.44), they were under the guardianship of their patron, and patrons as guardians had real power over a woman's legal acts (see Chapter 2, section V(f) above). The position of patronesses who had children was also improved by the Papian Act (G.3.49 - 53).

Gaius also had to deal separately with succession to the estates of freedmen who became Latins (G.3.55 - 73) (referred to in J.3.7.4), and capitulated aliens (G.3.74 - 76) (see Chapter 2, section I(c) above). The latter could be dealt with briefly, because they could not make a will and their patrons succeeded on intestacy, either as if the former slave had been free or as if he or she would have been a Latin had they not been disgraced. The position with regard to Latins was complicated by the Junian Act which gave the former slave's estate to the patron as if it were a slave's personal fund. The result was that it was originally a patron's heirs who had a claim if he predeceased and not the patron's children, who had a separate claim if the former slave became a citizen (see G.3.57 - 62, explaining the difference in regime). The position was further complicated by the Largian Resolution (AD 42) referred to in J.3.7.4, which provided that the patron had first claim, then his descendants who had not been disinherited, and only then his heirs (G.3.63); this gave rise to considerable difficulties of interpretation which are discussed in G.3.63 - 71. A Latin might also be given citizenship by imperial grant without prejudice to his patron's rights; his estate was then treated as if he had died a Latin unless he qualified for citizenship in some other way, for example by marrying under the Aelian-Sentian Act and producing a child (G.3.72 - 73).

Justinian already simplified the law when he provided that all slaves freed should become citizens (J.3.7.4; J.1.5.3). However, he then reformed the law on the estates of freedmen by the comprehensive pronouncement which is summarised in J.3.7.3 (C.6.4.4 of AD 531). He further altered the law by Nov.78.1 (AD 539), giving all freedmen rights as if they were *equites* and had been free-born, without prejudice to the rights of the patron, unless the patron chose to give them up.

Assigning freedmen (J.3.8). The institution of assigning freedmen to a particular descendant or descendants of a patron, depriving the others of their independent right to the succession under the *Twelve Tables*, is not mentioned by Gaius although it had existed from the time of Claudius (J.3.8.3). Although emancipation impliedly revoked the assignment (J.3.8.2), assignment could be made to an emancipated child and disinheriting a child did not affect the assignment unless intended to do so (Ulpian, D.38.4.1.6).

(c) Estate-possession
(J.3.9)

The topic of estate-possession spans both testate and intestate succession, although dealt with by Justinian as an appendix to the law of intestate succession (J.3.9 pr). In one sense it is an independent form of universal succession created by the praetor, who could not give a right of succession on death as heir directly (J.3.9.2). In Justinian's law the retention of estate-possession is a reflection of the historical growth of the law and the difficulty of fusing completely the state law, praetorian law, and imperial innovations. The position may be compared with that of equity and common law in English law (see Chapter 6, section V below). The existence of separate remedies in a law largely based on remedies was a major reason for the difficulty in both instances.

Although J.3.9.1 - 2 is largely taken from G.2.32 - 34, the remainder of the title gives details not found in Gaius' *Institutes* and the title gathers together various points on the working of estate-possession which are assumed in earlier references to it. By the time of Justinian a claim to estate-possession could be made quite informally (see J.3.9.12(7), which appears to refer to C.6.9.8 and 9 (AD 320 or 326 and 339)), but in classical law a petition to the praetor was needed (see J.3.9.9(8) - 11(6)) and a grant did not necessarily mean that the estate-possessor would retain the estate. The grant could be provisional (*sine re*) (G.3.35 - 38), meaning that the successor by state law could recover the estate from the estate-possessor if he or she chose to do so. If the estate-possession was *cum re* the estate-possessor could recover from the successor by state law or resist a claim made by the successor by state law; the praetorian law therefore superseded the state law in such a case. There was no longer this conflict in Justinian's law, where estate-possession simply extends the range of remedies available. Apart from the grant of estate-possession in ordinary course, the praetor could grant it at his discretion after investigation (*bonorum possessio decretalis*). An example was application on behalf of a person unborn (Ulpian, D.37.9.1 pr).

The remedies available to the estate-possessor are not discussed, but he had a special interdict to recover the estate in the first place, the interdict *quorum bonorum* referred to in J.4.15.3. After Hadrian it was available even against someone who had usucapted a vacant inheritance (G.2.57) (see Chapter 3, section II(c) above, and Chapter 6, section I below). The estate-possessor also had an adapted version of the heir's remedy, the claim for an inheritance (referred to in J.4.6.28).

(d) Justinian's scheme of intestate succession in the Novels

Some nine years after the revision made in the course of compiling the *Institutes*, *Digest*, and *Codex*, Justinian completely recast the law of intestate succession in Nov.118 (AD 543) (with a further amendment in AD 548 by Nov.127). In this new scheme the ruling principle is cognatic relationship; all distinctions between male and female and between agnate and cognate were removed (Nov.118.4), but the new scheme applied only to those who followed the orthodox catholic faith, not to heretics (Nov.118.6). The scheme also applied to statutory guardianship, but there there was a preference for males. Only a mother or grandmother among females could become statutory guardian, and in order to do so she must renounce any second marriage and the benefit of the Velleian Resolution (Nov.118.5).

The order of succession established was:

(1) Descendants, male or female and through male or female, regardless of whether they were within *potestas*. Division of the estate was by stems (*per stirpes*).
(2) Ascendants, through father or mother, with half the estate going to each branch if they were in the same degree of relationship and all going to the nearest if they were not; but, if there were brothers or sisters of the full blood, they shared equally with ascendants.
(3) Collaterals, with brothers and sisters of the full blood taking in preference to brothers and sisters of the half blood. Children of predeceasing brothers and sisters represented their parents, so that,

for example, children of a full brother would exclude a half brother. No other representation was allowed in competition with ascendants by Nov.118, but by Nov.127.1 children of full brothers could represent their parents when called along with other brothers and ascendants.

(4) Other collaterals more remote than brothers and sisters by degree and with division by heads, with no representation. No limitation was placed on the degrees of relationship within which a claim could be made.

No mention is made of husband and wife, who could claim in the absence of relatives under the older law, but it is assumed that they could still claim, as their right is not abolished (see *Basilica* 45.5).[33] By Nov.53.6 (AD 537) a widow or widower who was in need and had no dowry or antenuptial gift was given a right to a quarter of the deceased spouse's estate. This right applied even if the deceased spouse had left a will, but then any legacy given counted towards the quarter. By Nov.117.5 (AD 542) the position of the needy widow was modified by giving her a right to a quarter of her husband's property only if there were no children. If there were children, her right was reduced to a usufruct and her share became a quarter or an equal share with the children if there were more than three. It is commonly held that the needy widower was deprived of his quarter by the same enactment, but this interpretation of the reference made to widowers in the *Novel* has been challenged.[34]

If there were no other claimants the Fisc, the imperial treasury, claimed the estate.

IV. Other Cases of Universal Succession
(J.3.10 - 12)

In one sense estate-possession is not a case of universal succession comparable with succession by inheritance, because the position of the estate-possessor was created by the praetor and not by state law, but its connection with succession on death justifies treating it along with inheritance. Looked at from another point of view, it is a further example of universal succession to be grouped with the other miscellaneous cases dealt with or referred to in book 3, titles 10 - 12 of the *Institutes*. None of these cases compared in importance with succession on death, and brief comment on the brief titles suffices.

(a) Acquisition through adrogation
(J.3.10)

Adrogation is one of the two cases of universal succession other than succession on death which survive in Justinian's law, but there were significant differences from the position in classical law as described by

Gaius in G.3.82 - 84, which forms the parallel to J.3.10 pr - 1, 3. Gaius says that obligations created by the oaths of freedmen are destroyed and so are obligations arising from joinder of issue in a statutory court (G.4.103 - 109). The variations here reflect changes in the sources of obligations and in procedure. In the time of Gaius a usufruct was lost by any status-loss, although in practice the benefit could be continued by provision for re-creation of the usufruct in case of status-loss by change of family; Justinian changed the law by C.3.33.16 (AD 530). Obligations owed by the person adrogated were destroyed under state law because the person owing them ceased to exist; the classical remedy was policy actions granted by the praetor to the same effect as stated in J.3.10.3. The reference to usufruct in the last sentence of paragraph 3 refers to the fact that the person adrogated still had the ownership of the property in Justinian's law but his father had the usufruct (J.2.9.1); he would have had the enjoyment or usufruct as well if he had stayed independent.

The changes made by Justinian in the rights of the head of the family over the property acquired by members of the family (J.2.9.1 - 2) are reflected in J.3.10.2. The prior claimants to whom reference is made are set out in C.6.59.11 (AD 529): descendants and brothers from the same marriage or from another marriage where property was acquired from that marriage or from the mother. What the father keeps automatically is anything acquired through what he himself has given to the person under his authority (J.2.9.1).

(b) Assignment to preserve freedom
(J.3.11)

If no heir accepted an inheritance given by will, the will failed and with it testamentary grants of freedom. The institution of assignment to preserve freedom was a creation of the emperor Marcus Aurelius (AD 161-80) designed to preserve grants of freedom by providing a successor who would make an arrangement with creditors and allow the grants of freedom to stand where there was no intestate heir who would be prejudiced (J.3.11.4). If the estate were simply sold by the creditors the slaves would be part of the assets available to the creditors. Assignment applied only where an estate was bankrupt and liable to sale by creditors (J.3.11.2), not when it was claimed by the Fisc as ownerless; but grants of freedom were still given effect if the Fisc claimed (J.3.11.1; Papinian, D.40.4.50 pr). The pronouncement in J.3.11.1 is not preserved in the *Codex* although it is referred to in C.7.2.6 (AD 238-44). The assignee or assignees had to give security for payment of the creditors (J.3.11.1) and then took over the estate and apparently were treated as heirs (cf. C.7.2.15.1a of AD 531-2). From C.7.2.15 (AD 531-2) it appears that it was not necessary to offer to honour all the grants of freedom.

The pronouncement referred to in J.3.11.7 is C.7.2.15 (AD 531-2). From this it appears that the estate could be granted to one of the slaves (even one not given his freedom: C.7.2.15.5) as well as to an outsider (C.7.2.15 pr), and Justinian provided that the assignment was still possible within a year of sale of the assets. The pronouncement still applied if the creditors agreed to take less than full settlement. Slaves were not compelled to accept their freedom. If the estate was sufficient to pay the debts, all slaves freed were given their freedom even if the person taking the estate had not promised to give freedom to all; when there were several claimants they shared the estate if they claimed together, but if they claimed separately the first to offer full security was preferred. In the absence of such an offer they were taken in order (all within a year), but a preference was given to the one offering to make the greater number of grants of freedom (without prejudice to grants of freedom already made before the better offer was made within the year allowed).

(c) Obsolete cases of universal succession
(J.3.12)

Sales in Execution. The process of sale in execution or sale of a bankrupt estate is described in G.3.77 - 81. It applied, for example, where execution of a judicial decree was sought, where a defender failed to defend proceedings, or where a deceased debtor's estate was not claimed by a successor. It was also used where a debtor surrendered his estate to his creditors under the Julian Act of the early Empire, which allowed the debtor to escape personal execution and infamy by doing so.

The procedure was that possession of the estate was granted by the praetor to any creditor who applied for it. Time was given to the debtor if alive, or to his successor if he were dead, to appear and satisfy the creditors. If no action was taken by or on behalf of the debtor the creditors were ordered to appear and appoint one of their number to carry through a sale of the estate as a whole to a purchaser who offered a dividend on the debts and took over the assets. Physical things became part of the estate of the purchaser, who then acquired full title by usucapion. Debts did not pass automatically either to or against the purchaser but the praetor granted policy actions to and against him (G.4.34 - 35). In these ways he became universal successor. Where there were reciprocal obligations between creditor and debtor, the purchaser could sue only for the balance (G.4.65 - 68).

Under the later procedure for execution of decrees and recovery of debts, creditors still obtained possession of the estate but the assets were sold individually, at the best price possible, to the extent necessary to satisfy the debts. The fuller account of the law referred to is in D.42.5 (but see also C.7.72 and especially C.7.72.10, a pronouncement of Justinian of AD 532). Justinian provided in C.7.72.10 that if only some creditors obtained possession, others could come in along with them within two years (or four if they lived in another province). If they did not come in within this period, the creditors who had possession and realised assets were required to seal up and deposit any surplus in the treasury of the local church. The surplus assets were then available to later creditors. The terminology of sale in execution was kept despite the changes in procedure. The later procedure is similar to the disposal of individual assets (*bonorum distractio*) prescribed by senate resolution (Gaius, D.27.10.5) for distinguished persons such as senators. In that case, however, a supervisor was appointed by the praetor or provincial governor to conduct the sale of such assets as were necessary. Surrender to bankruptcy, under the Julian Act, remained possible in Justinian's law (cf. J.4.6.40).

The Claudian Resolution. The Claudian Resolution (referred to in J.3.12.1) is mentioned briefly in G.1.84, 91, and 160, the last two texts relating to the case in J.3.12.l. Paul's *Sentences* 2.21A gives a fuller account, and 2.21A.17 suggests that a praetorian decree was needed after the warnings in order to effect the enslavement. We are not informed of the property consequences in the *Sentences* or in Gaius' *Institutes*. We do not therefore know whether the model was the purchaser of goods or adrogation. Debts would be cancelled under state law as a result of the change of status, so that policy actions would be needed to and against the new owner of the woman's estate.

Other cases. Certain cases of universal succession found in classical law were so far obsolete that Justinian had no need even to refer to them. These are entry into marital subordination, referred to by Gaius in G.3.84 along with adrogation, and assignment in court of an inheritance, the effect of which he describes in G.3.85 - 87 (and also in G.2.35 - 37).

The effect of entry into marital subordination was the same as adrogation, with the person obtaining power over the woman taking the place of the adrogator (see J.3.10 and section (a) above). As marital subordination was virtually obsolete in Gaius' time it is clear why Justinian found no need to refer to it.

By assignment in court, described by Gaius in G.2.24, a statutory heir on intestacy who had not yet accepted the inheritance could transfer the position of heir to a third party, who then became the universal successor of the deceased. If the heir attempted to transfer the inheritance after acceptance, he did transfer the physical assets, but at the same time he destroyed any debts due to the estate while remaining liable to pay the debts of the deceased (G.3.85). Debts could not be transferred by assignment in court (G.2.38) and he was treated as renouncing them, but he could not renounce the position of heir to the prejudice of creditors. A testamentary heir who attempted to assign the inheritance in court before acceptance achieved nothing; if he attempted it after acceptance the result was the same as in the case of the statutory heir (G.3.86). The statutory heir had a claim by law to the estate, perhaps in very early law some expectation of benefit, in a way which a testamentary heir created by the deceased had not. In the case of an immediate heir, who did not require to accept the inheritance, Gaius records a dispute between Sabinians and Proculians (G.3.87). The former held that an attempted assignment was ineffective; the latter that the effect was the same as an assignment after acceptance, which seems logical. Assignment in court fell out of use by the fourth century and this application of it disappeared also.

Select Bibliography

General

Beinart, B., 'Heir and Executor', *AJ* 3 (1960) p. 223.
Buckland, W.W. and McNair, A.D., *Roman Law and Common Law* 2nd ed. rev. F.H. Lawson (Cambridge, 1952) ch. 5.
Champlin, E., *Final Judgments: Duty and Emotion in Roman Wills 200 BC – AD 250* (Berkeley, 1991).
Cherry, D., 'Intestacy and the Roman Poor', *TvR* 64 (1996) p. 155.
Crook, J.A., *Law and Life of Rome* (London, 1967) pp. 118-38.
———, 'Intestacy in Roman Society', *Proceedings of the Cambridge Philosophical Society* (new ser.) 19 (1973) p. 38.
———, 'Women in Roman Succession', in B. Rawson (ed), *The Family in Ancient Rome: New Perspectives* (London, 1986) pp. 58-82.
Daube, D., 'The Preponderance of Intestacy at Rome', *Tul L Rev* 39 (1965) p. 253 (= D. Cohen and D. Simon (eds), *Collected Studies in Roman Law* II (Frankfurt-am-Main, 1991) p. 1087.
———, *Roman Law: Linguistic, Social and Philosophical Aspects* (Edinburgh, 1969) pp. 71-5.
Dixon, S., *The Roman Mother* (London, 1989) ch. 3.
Gardner, J.F., *Women in Roman Law and Society* (London, 1986) pp. 163-203.
Gardner, J.F. and Wiedemann, T. (eds), *The Roman Household: A Sourcebook* (London, 1991) section VI.
Johnston, D., *The Roman Law of Trusts* (Oxford, 1988).
Keenan, J.G., 'Tacitus, Roman Wills and Political Freedom', *Bulletin of the American Society of Papyrologists* 24 (1987) p. 1.
Saller, R.P., *Patriarchy, Property and Death in the Roman Family* (Cambridge, 1994).
Sirks, A.J.B., '*Sacra*, Succession and the *lex Voconia*', *Latomus* 53 (1994) p. 273.
Watson, A., *The Law of Succession in the Later Roman Republic* (Oxford, 1971).
Westrup, C.W., *Introduction to Early Roman Law* II ('Joint Family and Family Property'), III ('Patria Potestas') (London, 1934, 1939).

Testate Succession: Types of Will, Capacity

Buckland, W.W., *The Roman Law of Slavery* (Cambridge, 1908).
———, 'The Comitial Will', *LQR* 32 (1916) p. 97.
Baldwin, B., 'The Testamentum Porcelli', in C. Cosentini et al. (eds), *Studi in onore di C. Sanfilippo* I (Milan, 1982) p. 39.
Daube, D., *Roman Law: Linguistic, Social and Philosophical Aspects* (Edinburgh, 1969) pp. 76-81.

Duff, P.W., *Personality in Roman Private Law* (Cambridge, 1938; repr. New York, 1971).
Johnson, J.R., 'The Authenticity and Validity of Antony's Will', *Antiquité Classique* 47 (1978) p. 494.
Sitianni, F.A., 'Was Antony's Will Partially Forged?', *Antiquité Classique* 53 (1984) p. 236.
Syme, R., 'The Testamentum Dasumii: Some Novelties', *Chiron* 15 (1985) p. 41 (= Birley, R. (ed), *Roman Papers* V (Oxford, 1988) p. 521).
Tellegen, J.W., 'Captatio and crimen', *RIDA* (3rd ser.) 26 (1979) p. 387.
——, *The Roman Law of Succession in the Letters of Pliny the Younger* I (Zutphen, 1982).
Tellegen-Couperus, O.E., 'The Origin of "quando minus scriptum plus nuncupatum videtur" used by Diocletian in C.6.27.7', *RIDA* (3rd ser.) 27 (1980) p. 313.
——, *Testamentary Succession in the Constitutions of Diocletian* (Zutphen, 1982).
——, 'Livy and Gaius on the Making of Wills by Women', *BIDR* (3rd ser.) 27 (1985) p. 35.
Watson, A., 'D.28.5.45(44): An Unprincipled Decision on a Will', *IJ* (new ser.) 3 (1968) p. 377 (= *Studies in Roman Private Law* (London, 1991) p. 61).
——, 'Illogicality and Roman Law', *Israel Law Review* 7 (1972) p. 14 (= *Legal Origins and Legal Change* (London, 1991) p. 251).

Testate Succession: Heirs

Bergsma-van Krimpen, B., 'The Complications of a Double Role', *RIDA* (3rd ser.) 20 (1973) p. 275.
Buckland, W.W., '*Cretio* and Connected Topics', *TvR* 3 (1922) p. 239.
Crook, J., 'A Legal Point about Mark Antony's Will', *JRS* 47 (1957) p. 36.
Daube, D., '*Maior dividat, minor eligat*', *IURA* 14 (1963) p. 176 (= D. Cohen and D. Simon (eds), *Collected Studies in Roman Law* II (Frankfurt-am-Main, 1991) p. 1073).
Tellegen, J.W., 'Slaves, Substitutes and Sources', in F. Pastori et al. (eds), *Studi in onore di A. Biscardi* IV (Milan, 1983) p. 509.
Tellegen-Couperus, O., 'Manumission of Slaves and Collusion in Diocl. C.7.2.12.2', in F. Pastori et al. (eds), *Studi in onore di A. Biscardi* V (Milan, 1984) p. 207.
Wallace-Hadrill, A., 'Family and Inheritance in the Augustan Marriage Laws', *Proceedings of the Cambridge Philosophical Society* (new ser.) 27 (1981) p. 58.
Watson, A., 'The Identity of Sarapio, Socrates, Longus and Nilus in the Will of C. Longinus Castor', *IJ* (new ser.) 1 (1966) p. 313 (= *Studies in Roman Private Law* (London, 1991) p. 55).
Zimmermann, R., ' "Coniunctio verbis tantum". Accrual, the methods of joinder in a will and the rule against partial intestacy in Roman-Dutch and Roman Law', *ZSS* (rom. Abt.) 101 (1984) p. 234.

Legacies, Trusts, and Codicils

Beinart, B., 'Fideicommissum and Modus', *AJ* 10 (1968) p. 157.
Daube, D., 'Sale of Inheritance and Merger of Rights', *ZSS* (rom. Abt.) 74 (1957) p. 234 (D. Cohen and D. Simon (eds), *Collected Studies in Roman Law* I (Frankfurt-am-Main, 1991) p. 649).

———, *Roman Law: Linguistic, Social and Philosophical Aspects* (Edinburgh, 1969) pp. 96-102.
———, 'A Commentary on D.36.2.26.1', in D. Cohen and D. Simon (eds), *Collected Studies in Roman Law* I (Frankfurt-am-Main, 1991) p. 541.
Johnston, D.E.L., 'Prohibitions and Perpetuities: Family Settlements in Roman Law', *ZSS* (rom. Abt.) 102 (1985) p. 291.
———, 'Munificence and *Municipia*', *JRS* 75 (1985) p. 105.
———, 'Trusts and Tombs', *ZPE* 72 (1988) p. 81.
———, *The Roman Law of Trusts* (Oxford, 1988).
Kehoe, D., and Peachin, M., 'Testamentary Trouble and an Imperial Rescript from Bithynia', *ZPE* 86 (1991) p. 155.
Plescia, J., 'The Development of the Juristic Personality in Roman Law', in C. Cosentini et al. (eds), *Studi in onore di C. Sanfilippo* I (Milan, 1982) p. 485.
Rickett, C.E.F., 'Charitable Giving in English and Roman Law. A Comparison of Method', *CLJ* 38 (1979) p. 118.
Rodger, A., 'D.35.2.2', *ZSS* (rom. Abt.) 89 (1972) p. 344.
———, 'The Paligenesia of Digest 36.2.13', *ZSS* (rom. Abt.) 89 (1981) p. 366.
Schanze, E., 'Interpretation of Wills: an Essay Critical and Comparative', in D.L. Carey Miller and D.W. Meyers (eds), *Comparative and Historical Essays in Scots Law: a Tribute to Professor Sir Thomas Smith* (Edinburgh, 1992) p. 104.
Tellegen, J.W., and Tellegen-Couperus, O., 'Joint Usufruct in Cicero's *Pro Caecina*', in P. Birks (ed), *New Perspectives in the Roman Law of Property. Essays for Barry Nicholas* (Oxford, 1989) p. 195.
Thomas, J.A.C., 'Perpetuities and Fideicommissary Substitutions', *RIDA* (3rd ser.) 5 (1958) p. 571.
Torrent, A., 'The Nature of the Fideicommissum "*si sine liberis decesserit*" ', *TvR* 43 (1975) p. 73.
Watson, A., 'Narrow, Rigid and Literal Interpretation in the Later Roman Republic', *TvR* 37 (1969) p. 351 (= *Legal Origins and Legal Change* (London, 1991) p. 27).
———, 'The Early History of *fideicommissa*', *Index* 1 (1970) p. 178 (= *Legal Origins and Legal Change* (London, 1991) p. 181.
Wolff, H.J., 'The *lex Cornelia de captivis* and the Roman Law of Succession', *TvR* 17 (1941) p. 136.

Intestate Succession on Death

Daube, D., 'Intestatus', *RHDFE* (4th ser.) 15 (1936) p. 341 (= D. Cohen and D. Simon (eds), *Collected Studies in Roman Law* I (Frankfurt-am-Main, 1991) p. 1).
Feenstra, R., 'The "Poor Widower" in Justinian's Legislation', in P.G. Stein and A.D.E. Lewis (eds), *Studies in Justinian's Institutes in memory of J.A.C. Thomas* (London, 1983) p. 39.
Hopkins, K., *Death and Renewal* (Cambridge, 1983).
Marshall, A.J., 'The Case of Valeria: An Inheritance Dispute in Roman Asia', *CQ* (new ser.) 25 (1975) p. 82.
Yaron, R., 'Two Notes on Intestate Succession', *TvR* 25 (1957) p. 385.

Notes

1. P. de Francisci, *Il trasferimento della proprietà* (Padua, 1924) presents this view.

2. On the early *consortium* between joint immediate heirs see G. MacCormack, 'Hausgemeinschaft and Consortium', *Zeitschrift für vergleichende Rechtswissenschaft* 76 (1977) p. 1; J.W. Tellegen, 'Was there a consortium in Pliny's Letter VIII.18?', *RIDA* (3rd ser.) 27 (1980) p. 295; and Chapter 5, section V(c) on partnership.

3. A. Watson, *Succession in the Later Roman Republic* (Oxford, 1971) pp. 4-7.

4. There is a possible exception in the case of the immediate and compulsory heir. See Kaser, *RPR* I p. 395.

5. P. Bonfante, *Corso di diritto romano* VI (repr. Milan, 1974) chs 4 and 5; see generally P. Voci, *Diritto ereditario romano* I (Milan, 1960) and II, 2nd edn (Milan, 1963); and *Studi di diritto romano* II (Padua, 1985) pp. 1-275.

6. Cf. J.M. Kelly, *Studies in the Civil Judicature of the Roman Republic* (Oxford, 1976) pp. 81-92; B.W. Frier, *The Rise of the Roman Jurists* (Princeton, 1985) pp. 37-8.

7. H. Maine, *Ancient Law, with Introduction and Notes by Sir F. Pollock* (London, 1906) p. 233; cf. also p. 238. He is criticised by D. Daube, 'The Preponderance of Intestacy at Rome', *Tul L Rev* 39 (1965) p. 253 (= D. Cohen and D. Simon (eds), *Collected Studies in Roman Law* I (Frankfurt-am-Main, 1991) p. 1087.

8. J.A. Crook, 'Intestacy in Roman Society', *Proceedings of the Cambridge Philosophical Society* (new ser.) 19 (1973) p. 38: see p. 39. But see D. Cherry, 'Intestacy and the Roman Poor', *TvR* 64 (1996) p. 155.

9. R. Martini, *Le definizioni dei giuristi romani* (Milan, 1966).

10. *Oxford Latin Dictionary* s.v. 'testamentum'; A. Ernout and A. Meillet, *Dictionnaire étymologique de la langue Latine*, 4th edn (Paris, 1967) s.v. 'testis'.

11. See e.g. P. Voci, 'Testamento pretorio', *Labeo* 13 (1967) p. 319 (= *Studi in onore di G. Grosso* I (Turin, 1968) p. 97) (= *Studi di diritto romano* I (Padua, 1985) p. 407, with appendix).

12. J.R. Johnson, 'A Note on Caesar's Military Will: Why was it a "Concessio Temporalis"?', *Labeo* 26 (1980) p. 335; V. Scarano Ussani, 'Il "testamentum militis" nell'età di Nerva e Traiano', in *Sodalitas. Scritti in onore di A. Guarino* III (Naples, 1984) p. 1383.

13. O.E. Tellegen-Couperus, 'Livy and Gaius on the Making of Wills by Women', *BIDR* (3rd ser.) 27 (1985) p. 359.

14. For discussion of conceptions of usufruct in Roman law see M. Bretone, *La nozione romana di usufrutto* I (Naples, 1962) and II (Naples, 1967).

15. H.F. Jolowicz, *Roman Foundations of Modern Law* (Oxford, 1957) pp. 108-11.

16. G. MacCormack, 'Impossible Conditions in Wills', *RIDA* (3rd ser.) 21 (1974) p. 263.

17. J.A.C. Thomas, 'Fictitious Satisfaction and Conditional Sales in Roman Law', *IJ* (new ser.) 1 (1966) p. 116.

18. D. Johnston, *The Roman Law of Trusts* (Oxford, 1988) pp. 109-16.

19. J.W. Tellegen, *The Roman Law of Succession in the Letters of Pliny the Younger* I (Zutphen, 1982) pp. 69ff.

20. See e.g. W. Ullmann, *Law and Politics in the Middle Ages* (London, 1975) pp. 56ff. The Ulpianic passage shows that the text originally referred to the Augustan marriage laws.

4. Succession

21. See J.W. Tellegen, op. cit. above note 19, pp. 83ff; see also pp. 110ff on Pliny, *Letters*, VI.33.

22. O.E. Tellegen-Couperus, 'Some Remarks concerning the Legal Consequences of the *querela inofficiosae donationis*', *RIDA* (3rd ser.) 26 (1979) p. 399.

23. J.A.C. Thomas, '*Rei hereditariae furtum non fit*', *TvR* 36 (1968) p. 489; G. MacCormack, '*Usucapio pro herede, res hereditariae and furtum*', *RIDA* (3rd ser.) 25 (1978) p. 293.

24. See P.W. Duff, *Personality in Roman Private Law* (Cambridge, 1938; repr. New York, 1971) pp. 162-7.

25. F. Wieacker, 'The *causa Curiana* and Contemporary Roman Jurisprudence', *IJ* (new ser.) 2 (1967) p. 151; J.W. Tellegen, '*Oratores, Iurisprudentes* and the *Causa Curiana*', *RIDA* (3rd ser.) 30 (1983) p. 293.

26. See e.g. F. Wubbe, 'Der Wille des Erblassers bei Iav. D.33.100.1', in H.-P. Benöhr et al. (eds), *Iuris Professio. Festgabe für M. Kaser* (Wien, 1986) p. 371.

27. See R.S. Rogers, 'The Roman Emperors as Heirs and Legatees', *Transactions of the American Philological Association* 78 (1947) p. 140; F. Millar, *The Emperor in the Roman World* (London, 1977) pp. 153ff.

28. On the Pegasian Resolution see U. Manthe, *Das Senatusconsultum Pegasianum* (Berlin, 1989).

29. See Johnston, loc. cit. above note 18.

30. H. Ankum, 'Quelques remarques sur le *fideicommissum* d'un usufruit légué dans le droit romain classique', *RIDA* (3rd ser.) 24 (1977) p. 133.

31. M. Meinhart, *Die Senatus consulta Tertullianum und Orfitianum in ihrer Bedeutung für das klassische römische Erbrecht* (Graz, 1967).

32. On the rights of patrons see A.M. Duff, *Freedmen in the Early Roman Empire* (Oxford, 1928) pp. 36ff, especially pp. 43-4; S. Treggiari, *Roman Freedmen during the Late Republic* (Oxford, 1969) pp. 68ff, especially pp. 78-80 and 81; on the Augustan legislation generally see P. Csillag, *The Augustan Laws on Family Relations* (Budapest, 1976).

33. R. Feenstra, 'The "Poor Widower" in Justinian's Legislation', in P.G. Stein and A.D.E. Lewis (eds), *Studies in Justinian's Institutes in memory of J.A.C. Thomas* (London, 1983) p. 39n2.

34. See Feenstra, loc. cit. above note 33.

5. Obligations

Robin Evans-Jones
Geoffrey MacCormack

I. Obligations
(J.3.13)

One thinks of an obligation in modern law as a duty to pay or do something imposed on an individual by a rule of law. In Roman law the perspective was somewhat broader, in that an obligation was conceived as a legal relationship between two individuals under which one acquired a right and the other the correlative duty. From this point of view an obligation might be regarded as an asset, an incorporeal thing, the treatment of which properly fell under the law of property. Indeed both Gaius (G.2.14) and Justinian (J.2.2.2) treat obligations as part of the law of things, that is, of property (whether corporeal or incorporeal) which possessed an economic value. This blurs the distinction between the law of property, as concerned fundamentally with real rights, and the law of obligations, as concerned fundamentally with personal rights. Certainly Justinian places less emphasis than Gaius on the nature of obligations as things, though his definition, which is probably taken from a classical source, specifies only the elements of duty and compulsion, not those of right and entitlement.

As regards the classification of obligations there are differences between Gaius and Justinian. Gaius (G.3.88, 89) is concerned with obligations recognised by the state law. These Gaius classifies according to the two main sources of obligation: contract and delict (wrong). Obligations arising from wrongs may all be regarded as of the same kind, but obligations arising from contract are further differentiated according to their particular source. Yet, the classification in Gaius' *Institutes* according to source is not exhaustive, since no mention is made of obligations arising as though from a contract or as though from delict. However, in *Golden Words* Gaius broadens the range of his classification and introduces the notion of 'various types of causes' (D.44.7.1 pr). Under this third source he includes obligations arising as though from contract and as though from delict. An indirect consequence of the introduction of 'various types of causes' as a source of obligation is that reference is now made to obligations recognised by praetorian as well as by state law.

Justinian's principal division of obligations is determined not by their source but by their nature. Obligations are classified as 'legal' or 'praetorian', that is, as belonging either to the state or magisterial law (see Chapter 6, section V below). Therefore, in the definition of obligation which he gives (J.3.13 pr), the phrase 'in accordance with the laws of our state' must be understood as referring to rules belonging to all branches of the law. His secondary classification is based upon the source from which the obligation arises. In this respect obligations are classified as arising from contract, as though from contract, from delict (wrong), or as though from delict. Justinian may here be following the systematic approach of *Golden Words*.

II. Obligations Contracted by Conduct
(J.3.14)

(a) Introduction

The four contracts contained in this title were gratuitous transactions concluded, in the main, between friends. The basis of the classification, which is the same in the *Institutes* of Gaius and Justinian, is that the obligation arose by conduct. The conduct in each case was the handing over of an object from one party to the other. The contracts were distinguishable from each other on the basis of the accompanying agreement of the parties concerning the purpose of the handing over. For example, if we have agreed that I shall look after your horse we conclude a contract of deposit, but if the arrangement is that I am to use it the contract is *commodatum*. Although for developed law in all instances it was the agreement which in practice was the ground of the obligation, this was not by itself the legally operative factor, and there was therefore no contract until the horse was actually handed over. The reason was that the principal obligation to which these transactions gave rise was to return the same object, or its equivalent, when asked. Clearly this obligation could arise only when and if the object was in fact first delivered.

The one example of contracts concluded by conduct given by Gaius is *mutuum* (G.3.90). He adds a further case of an obligation that arises by conduct – namely, the obligation to give back what has been paid in error – even though, as he himself notes, it does not proceed from contract or agreement (G.3.91). The explanation for this odd arrangement is that Gaius was following an old model in which the classification was of obligation to repay grounded on the handing over of property. This basis is shown by the fact that the appropriate remedy in each case was the action of debt, since the delivery had effected a transfer of ownership.[1] Gaius keeps this classification in the *Institutes* even though its basis had already changed to that of agreements concluded by the handing over of property, in which the position of payment in error was, of course, anomalous. In *Golden Words* Gaius deletes all mention of payment in error from the discussion of contracts concluded by conduct (D.44.7.1.3 - 6). However, albeit Justinian postpones its proper treatment to the title on obligations as though from contract, under the influence of Gaius' *Institutes* he also illogically incorporates a reference to payment in error in this context.

The difference in Justinian's account lies in the fact that, beside *mutuum*, he includes the other three contracts which become binding by conduct: *commodatum*, deposit, and pledge. Why should these have been omitted from the *Institutes* of Gaius? Sometimes it is argued that it was because none was recognised as a contract in his time. This was either because a contract was only a transaction recognised by the state law, and these as yet were only equipped with praetorian actions on the case, or because the source of the obligation sanctioned by the actions on the case was not contract but delict, the failure to return the property in each case being treated as a wrong.[2] In fact all four transactions gave rise to state-law contractual actions by Gaius' time and, in *Golden Words*, he himself includes them all in the account of contracts made binding by conduct. The explanation for the omission must therefore have been either that for brevity's sake he mentioned only *mutuum* or that the other three did not fit into the old classification of obligations concluded by conduct as described above: none entailed a transfer of ownership nor gave rise to the action of debt.

(b) *Mutuum*
(J.3.14 pr)

Mutuum was the earliest of the contracts binding by conduct to receive legal recognition, having been known at the time of the *Twelve Tables*. It was a loan of property that is consumed by use, especially money, but also food, drink, oil, corn, and the like. The major consequence was that the borrower acquired ownership of what was lent, the obligation being to restore the equivalent at the agreed time. The borrower's liability was absolute, as distinct from strict, in that his obligation to restore was not discharged even if he lost the property through an act of God, such as shipwreck. Special rules applied in the case of loans of money to persons engaged in overseas trade or to professional athletes to help further their training.[3]

The contract was gratuitous in the sense that it could not provide for the payment of interest. This rule probably derives from the origin of *mutuum* as a transaction between friends. If the lender wished to secure interest he had to obtain a stipulation from the borrower by which the latter promised repayment of both principal and interest. This in effect permitted the creditor to base his action for return of the principal sum either on the *mutuum* or on the stipulation. The law limited the maximum rate of interest to 12%, reduced by Justinian himself to 6%.

An important rule was introduced in about AD 50 by the Macedonian Resolution of the Senate.[4] Its object was to prevent sons within authority from borrowing on the strength of their expectation of inheriting money and succumbing to the temptation of hastening the death of their father. A person who lent money to a son within authority was prevented, by means of a special plea in defence, from bringing an action to recover it even after the death of the father.

Strictly the resolution applied only to loans of money and not, for example, to credit sales, even though the son in such a case might have promised heavy interest. Thus where money was advanced and a stipulation taken for payment of principal and interest, the resolution applied even where the action was brought on the stipulation and not on the *mutuum*, but it did not apply where money was promised by stipulation without originally having been advanced by way of *mutuum*. If the creditor had reasonable grounds for believing that the person to whom he advanced the money was not under authority the resolution did not apply.

(c) *Commodatum*
(J.3.14.2)

Commodatum was a loan of a non-fungible, such as a horse, for a particular use. The nature of the use had to be specified in the contract or be clearly inferable from it. Exceptionally the loan could be of a fungible where it was simply to be used for display (Ulpian, D.13.6.3.6; Gaius, D.h.t.4). It became accepted in time that land could also be the object of *commodatum*. The difficulty was that as a general rule the borrower was not entitled to the fruits of the object of a loan for use, so where it was land the practice was to treat the arrangement as *precarium*, an institution in which the 'borrower' had general use of the land and kept the fruits.

The principal obligation in *commodatum* was that of the borrower to return the selfsame object at the end of the loan. If no period was agreed, he could keep it for a time reasonable for such a loan. To enforce the obligation, from at least the late Republic, the praetor provided an action on the case, but before the time of Gaius this was supplemented and ultimately replaced by a state-law action. It is likely, but not certain, that this contained a good-faith clause. Damages in classical law comprised the lender's loss, which would usually be the value of the object. Once he had handed over the object the lender was obliged to let the borrower keep it for the agreed time, premature repossession without good cause giving rise to the counter-action of loan for use (Paul, D.13.6.17.3). The lender was also liable for such expense incurred in the use of the object as would not reasonably be entailed by its normal use, and for loss suffered by the borrower through his, the lender's, intentional fault: for example, where he had knowingly lent faulty storage jars and the wine put in them by the borrower was lost or spoilt (Gaius, D.13.6.18.3). In the event of expenses having been incurred, the borrower was entitled to retain the object until they were paid (Paul, D.47.2.15.2).

Commodatum was gratuitous; if a fee was paid it was hire. Also, normally it was exclusively in the interest of the borrower, hence his standard of liability was high. J.3.14.2 explains that although, unlike *mutuum*, the borrower in *commodatum* was not liable for acts of God, he was nevertheless bound to show the highest standard of care in keeping the thing safe, and therefore was liable where loss or damage occurred through failure to maintain the degree of care expected of a responsible adult. The justification for the imposition of this standard of liability lay in the principle of utility; since the loan was purely in the interests of the borrower he should be bound to exercise all reasonable care with respect to the object. Should the borrower put the object to an improper use (Justinian's example is taking it out of the country), he incurs liability even with respect to acts of God. The position is generally thought to have been different in classical law when the borrower is said to have had an insurance liability (*custodia*), a standard which was stricter than that of fault. This is thought to have meant that although he was not liable for acts of God, he was liable for accidental loss of or damage to the object in circumstances where he had not in any way been at fault. The basis for such a view is the passage in the *Institutes* of Gaius (G.3.206)[5] which grounds the borrower's action for theft on his insurance liability. But Gaius does not explain what he means by the phrase 'insurance liability'. It remains possible that he was stating the borrower to be under an obligation to keep the object safe, and that where it was stolen the presumption would be that he had broken the obligation and was liable on the contract. Yet it might still have been open to the borrower to rebut this presumption by showing that he had exercised every possible care. Hence, the problem of the extent of the borrower's liability might involve consideration of the onus of proof. A borrower with the onus of proving absence of fault can be said to be under a 'higher' standard of care than one who does not have this onus. A further difficulty with the interpretation of the Gaian passage is that it is speaking only of loss suffered through theft. It is not clear that the same considerations applied where loss was otherwise caused.

If the borrower put the object to a use not authorised by the contract and which he had reason to believe the lender would not have approved, it constituted theft. In such circumstances the lender had both the action for theft and the action on the loan for the breach of contract.[6]

The borrower received only physical control, often called detention, of the object. As a result the lender need not be the owner; even a thief could make a valid *commodatum* of stolen property (Paul, D.13.6.15; Marcellus, D.h.t.16).

(d) Deposit
(J.3.14.3)

Deposit was the gratuitous contract whereby someone handed over a moveable object to another to be looked after. Its history was similar to that of *commodatum* with one important difference: a twofold pecuniary penalty was established by the *Twelve Tables* 'on the grounds of deposit'.[7] Generally it is thought that the action was available for all cases of deposit, but this is unlikely. The scope of the remedy was much narrower, probably introduced to sanction deposits at arm's length: those made with people with whom there had been no previous close relationship such as friendship. In particular an action would lie for a failure to return by a person, such as the custodian of a temple, who represented an institution which held itself out as a safe place to deposit goods.[8]

The normal case of deposit between friends was first enforced through the grant of an action on the case by the praetor in the late Republic. Before the time of Gaius, this was supplemented and later superseded by a state-law action for deposit which contained a good-faith clause. The principal obligation was that of the depositee to return the object when asked. However, he was also liable for any damage to it which he deliberately caused. In both cases the measure of damages was the depositor's loss. He, in turn, was liable in the counter-action on deposit for any expenses incurred by the depositee or any loss suffered by him due to his fault. Until he was paid, certainly in classical law, the depositee had a right to retain the object.[9]

Since deposit was exclusively in the interest of the depositor and gratuitous, the standard of liability of the depositee was low, being only for intentional fault. He was therefore not liable if the object was lost, damaged, or stolen through non-deliberate fault such as carelessness. However, the standard of liability could be increased by special agreement. In the event of being condemned for deliberate fault the depositee incurred infamy.

The purpose of the contract was safekeeping; hence any use of the object or indeed any handling of it by the depositee contrary to this purpose constituted not only a breach of contract, but theft. However, since the depositee received detention and not possession of the object, anyone, even a thief, could validly deposit goods.

Some special cases of deposit should be noted. First, there was emergency deposit, when the depositor, faced with a crisis, for example the collapse of his house or fire, had to act quickly and might not have been able to select his depositee with due care. Here the latter was liable for double the value of the deposit if he deliberately failed to restore it. This remedy was the old action of the *Twelve Tables* given a new scope by the praetor (Ulpian, D.16.3.1.1). Second, *sequestratio* was a deposit made jointly by two or more persons who were in dispute as to who was entitled to the object deposited (Ulpian, D.16.3.5). Pending the decision of the court the object might be entrusted to a neutral third party (*sequester*) who was bound to deliver it to the party to whom it was awarded by the court. The most important difference from ordinary deposit was that the *sequester* received legally protected possession, not just detention, of the object. This prevented usucapion from running in favour of either depositor and also meant that the *sequester* was entitled to the possessory interdicts. Third, the so-called 'irregular' deposit primarily concerned deposits of money upon the terms that the recipient – often but not necessarily a banker – was bound to restore not the same coins but an equivalent, ownership of the deposit being transferred to him. As he was owner the depositee could use the money, in which case it was usual to pay interest on it to the depositor. There are clearly similarities between this contract and *mutuum*. However, they were different: the depositee's primary intention was that his money be looked after; the arrangement was mainly in his interest. On the other hand a *mutuum* was given to someone in need and was in his interest. Thomas draws an analogy with modern banking practice: payment into a current account is like an irregular deposit; obtaining an overdraft like receiving a *mutuum*.[10] The main practical consequence of the distinction was that in deposit the payment of interest could be agreed upon by pact and a single action on deposit lay to recover both principal sum and interest. It used to be assumed that irregular deposit was a post-classical or Byzantine institution, but now the balance of opinion is that it was classical in origin. One may note that the action on the case cannot ever have applied to irregular deposit since it specified the return of the same object deposited.

(e) Real security
(J.3.14.4)

The earliest form of real security involved the transfer of ownership of an object capable of mancipation to the creditor with an ancillary agreement (*fiducia*) that the property be remancipated on repayment of the debt (see Chapter 3, section IV(d) above). This was not a satisfactory arrangement from the debtor's point of view since he lost both the use and the ownership of his property. In the late Republic a different form of real security called pledge (*pignus*) grew up alongside trust. The contract became binding on delivery of the pledge to the creditor, and he acquired not ownership but only possession. The principal obligation was that of the creditor to return the pledge unimpaired once the debt had been paid. As in *commodatum* and deposit, it was enforced by a praetorian action on the case which was supplemented and later superseded by a good-faith state-law action on pledge. The creditor was not allowed to use the pledge or to make a profit out of it, any acquisitions having to be set against interest and debt (Ulpian, D.13.7.22 pr), subject to the qualification that in classical law, in an arrangement called *antichresis*, the creditor took the fruits of the pledge in lieu of interest (Marcian, D.20.1.11.1). The creditor could sell the pledge if the debt was not paid on time. At first this right had to be specially agreed, but after the time of Gaius it was implied in the contract and could only be excluded by special agreement (G.2.64). The debtor was liable in the counter-action on pledge to reimburse the creditor for expenses properly incurred in the care of the pledge and for any loss he had suffered from it due to the former's fault. The action also lay if the object did not constitute a security, as, for example, where the debtor's title was defective, whether he knew it or not (Ulpian, D.13.7.9 pr). In addition to the action on the contract, the creditor was entitled to the possessory interdicts.

Like the borrower in *commodatum*, the creditor must show the highest standard of care in looking after the pledge (J.3.14.4). Some scholars have argued that in classical law he bore an insurance liability, but the likelihood is that both parties were always liable only for fault.

The various forms of real security are treated in further detail in Chapter 3, section IV(d) above.

III. Obligations by Words and their Applications
(J.3.15 - 20)

(a) Stipulations
(J.3.15)

The speaking of words, like the principle underlying obligations formed by conduct, was an extremely old method of contracting liability. The most important example of an obligation created by words, and indeed the only one to be mentioned by Justinian, was the contract called stipulation which was known to the *Twelve Tables*. Gaius mentions two other contracts in this context (G.3.95a - 97): the constitution of a dowry by formal promise, and the sworn promise by a freedman for the performance of services or the making of a gift to his patron. Both are examples of an obligation created by a unilateral promise, the promise in the case of the freedman being coupled with an oath. The freedman case survived into the law of Justinian (though it was not practically important), but the constitution of a dowry by formal promise had become obsolete.

The essence of stipulation was the exchange of an oral question and answer. Its other requirements varied at different periods. The very earliest form, accessible only to Roman citizens, was the use of the verb *spondere*, 'to promise solemnly'. By the time of Gaius other verbs were in use which were available to foreigners as well as citizens. Gaius mentions 'to give' (*dare*), 'to promise' (*promittere*), 'to promise faithfully' (*fidepromittere*), 'to authorise faithfully' (*fideiubere*), and 'to perform' (*facere*) (G.3.92). Although the matter is disputed, the likelihood is that these were simply the most common examples and not in fact the only verbs which could be used to conclude a stipulation at this time.[11] The principal further requirement in classical law was that the verb used in the answer must be the same as that used in the question; in other words there must be congruency between question and answer. If the question said, 'Do you promise?' the answer must say, 'I promise', and not, for example, 'I will do', or 'yes'. The answer had to follow the question without lapse of a lengthy period of time. Deliberation was expected to have been completed before the formal putting of the question. The final requirement was that the parties must exchange question and answer in each other's presence, not communicate, for example, by letter or messenger, though a party could be represented by a slave.

By the time of Justinian the position had changed, though both the steps in the process of change and the eventual result are a matter of controversy. The principal evidence is J.3.15.1, which summarises the history of the stipulation. After referring to the fact that certain verbs were once used – he gives those listed by Gaius – Justinian states that it does not matter which language is adopted or even whether both parties used the same language in question and answer; all that is necessary is for the content of the reply to be in accordance with that of the question. He then refers to the fact that the necessity for 'special words' was removed by a pronouncement of AD 472 of the emperor Leo (AD 457-74) which required only that the parties understand each other, whatever words they used. There are two ways of taking this passage. One may say that the pronouncement of Leo abolished the need for question and answer, leaving the position that any agreement was actionable as a stipulation provided it was not immoral, impossible, or illegal. On this view there would, in the law of Justinian, have been no difference between a stipulation and a pact. In view of the emphasis on question and answer still found in the *Institutes* itself and in the *Digest*, it is unlikely that so extreme a change was made by Leo. More probable is the fact that the pronouncement dispensed with the necessity for congruency but left intact the fundamental requirement of question and answer. Indeed it is possible that even before Leo the requirement that there be congruency between question and answer had been doubted. Leo's pronouncement may have had more a declaratory than an innovative force in that it simply settled these doubts. Consequently we assume that for the law of Justinian, question and answer were still needed but they need no longer be congruent. However, there must have been a spoken reply; a nod of the head did not count as an answer.

The question of formalities raises a further problem, that of the relation between the oral stipulation and the document in which it was frequently recorded. Strictly the document was always merely evidence that a stipulation had taken place after the observance of the correct oral formalities. Nevertheless there was a tendency for the document itself to be accepted as dispositive in the sense that what it recorded was assumed to have occurred, there being no consideration of whether the parties had in fact gone through the correct oral forms. Yet, since it did not create the obligation but was merely evidence of the contract, there remained the possibility of challenging the document on the ground that the oral requirements had not been observed. It is not clear whether the onus of proof was on the creditor to show that the oral requirements had been complied with, or whether it was on the debtor to show that, despite what was alleged in the document, some essential formality had been omitted.

A second problem concerned the interpretation of a document which recorded that someone owed something to another. What exactly did it have to say before the inference was made that the property was owed under a stipulation? In classical law the document had to record the fact of question and answer. This was certainly also sufficient for the time of Justinian. However, there was some relaxation in respect of elliptical statements which did not record the exchange of question and answer but said merely that A promised to pay B something, or even just that A acknowledged that he was indebted to B with respect to something. What was the extent of this relaxation?

In a pronouncement to the advocates of Caesarea, Justinian (J.3.19.12) addressed himself first of all to the problem of the dispositive effect of documents. He adverts to the problem of constant litigation caused by parties who alleged that they had not in fact made the oral contract which a document recorded them as having concluded. Consequently Justinian enacted that where a document stated that the parties to the contract were present, a presumption of presence arose which might be rebutted only by proof that one or the other of them was in a different place for the whole of the day on which the contract was said to have been made. The pronouncement also dealt with documents recording stipulations made by slaves and provided that it was not open to the owner to deny either ownership or the presence of the slave. One notes that the pronouncement addressed itself only to the presence of the parties. It did not touch on the question of what formalities, if any, the parties, when present, had to complete if the document was to raise the presumption that a stipulation had been concluded. Justinian deals with this matter at J.3.19.17, drawing upon a classical source, perhaps the *Sentences* or *Institutes* of Paul.[12] He says that where a document states that someone has promised something, the inference is to be drawn that the promise has been given in response to a preceding question. Such a statement provides evidence that no document could be used to prove a stipulation unless it was executed in a particular form. Although it need not actually say that one party put a question which the other accepted, the document must use language which allowed this to be inferred; that is to say, it must use the verb 'to promise' or one with an equivalent sense. Certain important consequences follow: a document which stated, not that A promised to pay 10 to B, but merely that he owed 10 to B, would be evidence not of a stipulation but of a *mutuum*. A document which stated that A and B had agreed that A would pay 10 to B would be evidence in itself neither of a stipulation nor of a *mutuum* but of an unenforceable pact.

To sum up the above discussion: in both classical and Justinianic law the document acted simply as evidence that a stipulation had been concluded. In classical law, to constitute evidence, the document had to be executed in a particular form: it must record the presence of the parties and the fact that a question and answer had been exchanged. In the time of Justinian the form was different: the document must still record the presence of the parties, but now it was sufficient to state that A had promised B, from which it was inferred that question and answer had been exchanged. So long as the parties had full capacity to contract and the document was executed correctly, it raised a presumption of stipulation which could be rebutted only by proof that one of them was in another place throughout the day on which the document recorded that the stipulation was concluded.

J.3.15.2 states that a stipulation is either absolute, postponed, or conditional. A postponed stipulation is one which contains a clause specifying that what was promised did not become due until a certain date; for example, I promise to pay 10 on the first of March. The date might be known, as in this example, or unknown, as where performance was due 'when X dies'. The obligation arose immediately on conclusion of the stipulation but could not successfully be sued upon until the specified time had elapsed. The debtor was entitled to discharge the obligation before the appointed date, and if he did so he could not subsequently reclaim the money as a payment which was not due (Pomponius, D.12.6.16.1).

Difficulties were caused by a different type of time clause, namely one which specified that what was promised was due only for a certain time, as where a promise was put in the form, 'Do you promise to pay 10 every year for so long as I live?' In classical law the rule was that a person could not be placed under an obligation limited by time.[13] Where a stipulation purporting to create such an obligation was concluded it was not held to be void, but the state law ignored the specification of time. The praetor, however, adverting to the intention of the parties, allowed the defence of fraud or contrary agreement if an action was brought after the agreed time. The same rule is given in J.3.15.3, which mentions only the plea of contrary agreement.

A stipulation might also be subject to a condition (J.3.15.4). In this case the very existence of the obligation was made to depend on some future uncertain event, as, for example, 'if Titius is made consul'. It is distinguishable from a postponed stipulation, where the date for performance was unknown, by the fact that in that case, although unknown, the date was nevertheless bound to occur. Since the obligation did not yet exist, payment made before the realisation of a condition could be reclaimed as not due (Pomponius, D.12.6.16 pr).

J.3.15.4 mentions that the promise to give 'if I never go up to the Capitol' was treated like a stipulation for a thing to be given to the creditor on his deathbed. This means that the obligation commenced in the last moment of the creditor's life when it was no longer possible for the terms of the condition to be broken. The remedy would lie to the creditor's heir.

A stipulation gave rise to a unilateral obligation of strict law (see Chapter 6, section VII below). The promisor was held liable strictly in accordance with the wording of his promise. The judge had no power to consider matters extraneous to its terms. Thus even where the stipulation had been extracted under circumstances constituting fraud or deceit, it remained valid and grounded an action. Only with the introduction by the praetor in 66 BC of the defence of fraud did it became open to a debtor, if sued upon such a promise, to plead that condemnation in the action would constitute a fraud on him. The onus was on the debtor to prove the occurrence of circumstances sufficient to make out fraud.

The promisee had two possible remedies (J.3.15 pr). If the stipulation was for a fixed thing or quantity he had the action of debt; if it was not for something fixed, for example, for the doing of a job, he had the action on a stipulation in which he sought his interest in the non-performance of the promise. In this case it might be difficult to quantify his actual interest or loss, hence a prudent creditor would require the promisor to add to his stipulation the promise of a penalty of a fixed sum of money to be paid if the primary promise was not fulfilled (J.3.15.7). The use of the stipulation with a penalty clause played a very important role in Roman law in securing the enforceability of arrangements that could not directly be enforced.[14]

(b) Multi-party stipulations
(J.3.16)

In this title Justinian discusses joint creditors and joint debtors. The contract by which such a result was achieved required the appendage of a second stipulation to a preceding stipulation, the second stipulation, at least in the case of the joint debtors, being accessory to the first. The additional stipulation in each case did not result in the formation of a second debt. The purpose of each institution was rather that a single obligation could be discharged by payment to, or demanded in full from, any one of a number of people. The others, if joint debtors, were released, or, if joint creditors, were barred from raising a second action in respect of the same debt.

Justinian's treatment of joint creditors differs from that of Gaius. Gaius gives a full account of an institution which he says was commonly known as adstipulation (G.3.110 - 117). The essence of the arrangement was the provision of a joint creditor by getting a third person, called the adstipulator, to stipulate for the 'same thing' as the main promisee. Note that if the debtor, instead of the 'same thing', promised, say, 10 in addition to the 10 promised in the main stipulation, the effect was to create two separate obligations for 10. Exceptionally the adstipulator might validly stipulate for 'less' than the principal promise but never for more.

Payment or other performance to the adstipulator discharged the obligation and the adstipulator, being a party to the contract, could sue for what was due.

The main use of adstipulation in classical law was for the case in which the promise itself infringed the principle that no right might begin in the person of the heir (G.3.117).[15] The rule that a stipulation was void which was intended for the benefit of the heir, and was to take effect after the death of the promisee, could be circumvented by use of adstipulation. The promise to the adstipulator was valid and he might enforce it, being liable to the main promisee's heir on the mandate given by the promisee.

Justinian himself says nothing about adstipulation, perhaps because its main function mentioned by Gaius had become obsolete now that obligations might begin in the person of the heir (J.3.19.13). Instead he considers the position where there are two or more principal beneficiaries of a promise. After they in turn have put the question to the debtor he makes the following single reply to them all: 'To each of you I promise to give X.' He must not reply individually to each question since he would thereby conclude a number of separate stipulations for X.

Just as joint creditors might be created by an additional stipulation so might joint debtors, an arrangement constituting the most important means for providing security in Roman law. The creditor in this case put the question to the principal debtor, 'Do you promise 5?'; and then he asked the other debtor, 'Do you promise the same 5?'. The debtors in turn replied 'I promise' to the question put to them.[16]

J.3.16.1 discusses the effect of contracts for joint creditors and debtors. Both arrangements gave rise to what is called solidary liability. By identifying what this entailed it is possible to distinguish the adstipulation of classical law from the position of joint creditors to which Justinian refers.

The essence of solidary liability is simple, its detail complex. Derived from the term *in solidum*, it means that an action lies against an individual debtor or can be brought by an individual creditor for the full amount of the debt in question. By contrast, individual co-heirs, for example, in respect of inherited rights, could bring their action, not for the full amount of the debt owed to the deceased, but only in proportion to their share of the inheritance.

The first difficulty is that there are two main sorts of solidary liability: (1) where several people were liable (or entitled) for a joint delict, each must pay (or recover) the full penalty and payment by one did not release the other joint wrongdoers; and (2) where each of two or more persons was liable or entitled to the whole debt, but it was due only once, with the consequence that if the sum was once paid the whole was ended.[17] The latter is what J.3.16.1 says of joint creditors and debtors. However, within this class of solidary liability there is a further subdivision to be made. In classical law the mere bringing of an action by or against one of the parties in the solidary relationship barred or released all the others. The reason was that joinder of issue between one of the creditors and one of the debtors destroyed the whole obligation. One major effect was that if I raised my action against one of two debtors and he turned out to be insolvent, I could not subsequently raise an action against the other debtor. This regime was given the name 'correality'. In J.3.16.1 the position is different. It is no longer joinder of issue but actual performance in full which extinguished the debt. Thus if I raise my action against an insolvent or part-solvent debtor I am no longer barred against the others. This is called 'simple solidarity'.

The relationship between correality and simple solidarity has usually been explained historically: the former was a classical institution and the latter Justinianic. However, it has become increasingly accepted that simple solidarity originated in classical law, probably in those actions which contained good-faith clauses in their formulae.[18] Certainly by Justinian's time it had replaced all cases of correality.

Joint debtors and joint creditors therefore stood in a relationship of liability which was simple solidarity in Justinianic law. How did this differ from the position of adstipulators? The answer is that the latter stood in a relationship very similar but not identical to solidarity. As in solidary liability, the adstipulator could discharge and sue for the debt in full. The difference lies in the relationship between him and his principal creditor on the one hand, and that between the creditors in Justinian's time on the other. The adstipulator was an accessory creditor, whereas in the Justinianic institution they were all joint principal creditors. In early law there was no contractual relationship between the adstipulator and the principal creditor, as shown by the fact that section two of the Aquilian Act provided a penal remedy against an adstipulator who released the debtor in fraud of his principal. However, by classical times there was a contractual relationship of mandate between them, under which the principal could recover anything received by the adstipulator from the debtor (G.3.215, 216).

A further feature of the solidary liability existing between the joint principal creditors was that there was no legal relationship between them arising from their stipulations, with the result that if one had recovered the debt in full the others could not claim their contribution from him. In practice the severity of this rule was mitigated by the fact that joint creditors often stood in a quite separate legal relationship, such as partnership, on the basis of which they would have had a measure of accountability to each other.

(c) Stipulations taken by slaves
(J.3.17)

Slaves, as human beings, had the natural capacity to engage in the act of concluding a stipulation, but they lacked the legal capacity to acquire for themselves any right or obligation. Under state law, any right for which the slave stipulated was acquired for his owner. For this purpose a vacant inheritance of which he formed part was deemed to be owner, but the stipulation would fail if the heir never entered. Obligations incurred by the stipulation of a slave could not be enforced under state law against the master, but, in the appropriate circumstances, they could be enforced by one of the praetorian remedies such as the action about the personal fund.

The rule stated in J.3.17.2 is not an exception. Where the slave stipulates for a right the content of which is an act to be performed personally by him, such as the right for him to cross the promisor's land, he is the only person whose act will qualify under the right. Nevertheless, the right itself passes to the owner in the sense that he has the action on the stipulation if it is infringed. One must distinguish a stipulation under which a right of way is created. This is acquired for the owner of the land in question and runs with the land.

J.3.17.3 places command on the same level as nomination. A co-owner is to acquire the whole either if the slave expressly takes in his name or if he is commanded so to acquire by that co-owner. In classical law the position with respect to command was disputed. Gaius reports that the Sabinians had given command the same legal effect as nomination, whereas the Proculians held that the command was to be treated as void (G.3.167a). Justinian here adopts the Sabinian view.

(d) Classification of stipulations
(J.3.18)

The essential distinction made by Justinian is that between stipulations voluntarily concluded by the parties and those which they are forced to conclude by a magistrate or judge. The former is in effect the ordinary verbal contract which might cover any subject matter within the limits permitted by law. The latter are either adjuncts of legal proceedings or mechanisms for the provision of some specific legal remedy. Justinian gives several examples of judicial and praetorian stipulations.

The undertaking (*cautio*) against fraud might be imposed by a judge in a case where the owner sued the possessor in good faith to protect the plaintiff against any possible adverse dealing with the property by the defendant (Gaius, D.6.1.18, 20; Ulpian, D.h.t.45).

The undertaking for pursuit of a fugitive slave or restoration of his price was exacted where a slave, claimed by the owner from a person who improperly had possession of him, had taken to flight (Ulpian, D.4.2.14.11), or where, through the fault of the heir, a slave left as a legacy had taken to flight (Ulpian, D.30.47.2). The clause on restitution of the price probably means that the defendant was to pay the value should he be unable to recover the fugitive, although the texts do not make the point entirely clear.

The undertaking against imminent loss might be ordered where a person was able to show that his property was threatened through the state of neighbouring premises, the owner of the latter being required to provide security for payment of compensation should the threatened damage actually occur (Ulpian, D.39.2.7 pr).

Under certain circumstances, as where a legacy was left subject to a condition, the heir was required to give security for its future payment if and when it became due (D.36.3). For the giving of security by guardians and supervisors see J.1.24, and Chapter 2, sections V(d) and (e) above.

The stipulation for ratification by one's principal was required where the prosecution of an action was conducted through a representative, who was ordered to give security that his principal would not subsequently sue again on the same matter (J.4.11 pr).

(e) Ineffective stipulations
(J.3.19)

In spite of the rubric, this title contains a number of sections which have nothing to do with the conditions under which stipulations were void. Thus, for example, J.3.19.17 deals with the effect to be given to a document which records that someone gave a promise. Even those parts which deal with ineffective stipulations are badly arranged.

A stipulation was formed by the oral exchange of question and answer. It therefore required the presence of the parties (J.3.19.12) who must be able to hear and speak (J.3.19.7). An insane person, except in a lucid spell, could not conclude this as any other contract, because he could not understand what he was doing. Similarly the position of children under guardianship was distinguished depending on whether they could understand what they were doing or not. In the time of Justinian, understanding was deemed to be acquired at seven years. Children below this age, like insane persons, were unable to stipulate (J.3.19.10). Those of seven and over could stipulate with the endorsement of their guardian if they were to be put under an obligation. If they were the beneficiary of the promise, they could stipulate even without the guardian's endorsement (J.3.19.9; cf. G.2.83). On the ground of convenience the law was more ready to accord recognition to the contractual acts of a child around the age of seven than to impose delictual liability on such a child (cf. J.4.1.18(20)). Hence Justinian, echoing Gaius, remarks on the 'generous interpretation of the law' in the matter of contract. But no such relaxation was granted in the case of children under the age of puberty still in the power of their father; even with his endorsement they were not permitted to bind themselves by contract.

A stipulation was invalid if the content of the answer differed from that of the question. One example is that where the promisor agrees to do what the promisee asks but qualifies his acceptance by the insertion of a condition. Another case caused difficulty. J.3.19.5, following Gaius (G.3.102), declared to be void a stipulation in which A asks B whether he will promise 10 and B replies that he promises 5 (or vice versa). Both sources here take the difference between 10 and 5 as disclosing lack of congruity between question and answer. In fact, at least for Justinian's time, this ruling was misleading. It is clear from a *Digest* passage that, possibly in the late classical law and certainly in the time of Justinian, such a stipulation was held valid with respect to the lesser sum, the reason being that there had been agreement on this (Ulpian, D.45.1.1.4).

The parties might seem to be in agreement when in fact they were not, as where both stipulated for the delivery of Stichus but one had in mind another slave whom he believed to be Stichus (J.3.19.23). This error in respect of the object of the contract made it void.

Also void were impossible stipulations. Impossibility might be a matter of fact or of law: it was of fact, for example, where the promise was to deliver a hippocentaur, a creature which cannot exist (J.3.19.1); of law, for instance, where the promise was to deliver an object which already belonged to the creditor (J.3.19.22) or to the state (J.3.19.2), since in neither case was it possible for the promisor to transfer ownership of the object to the promisee. Should a stipulation provide for the performance of something intrinsically possible but make it conditional upon an impossibility, whether physical or legal, the same result obtained and the whole stipulation was invalid. To some extent the effect of this rule was curtailed by the further rule that if such a condition were framed negatively, for example, 'if I do not touch the sky', it was ignored and the stipulation was valid, the negative formulation entailing that the condition could never operate to destroy the obligation (J.3.19.11).

Impossible conditions must be distinguished from those termed 'preposterous' because they produced an absurdity. A common case is that in which the promise is to be fulfilled before the condition can be realised: 'If the ship arrives from Asia sometime, do you promise to give today?' (J.3.19.14). Here the condition contemplates an event happening in the future and yet the promise specifies that the obligation is to be fulfilled today. In classical law preposterous conditions were treated in the same way as impossible conditions and the whole stipulation held void. Justinian, however, extending the terms of an earlier pronouncement of Leo, provided that the condition should be so interpreted as to remove the absurdity, as by deletion of the word 'today' in the above example (C.6.23.25 of AD 528).

Stipulations for the performance of an illegal act, for example, to commit murder (J.3.19.24), or stipulations for something regarded as contrary to good morals, for example, 'make me your heir', were void. Also void were promises of payment conditional upon such illegal or immoral acts.

Where a party to a stipulation died, in the normal case his right or obligation was inherited by his heir. But where the promise was for a personal service, for example, to paint the promisee's portrait, death of either party discharged the stipulation. Difficulty was caused by the promise of an act which did not constitute a personal service, for example, to build a shelter for the poor. In early law such acts were discharged by the death of the party bound to do the act. Eventually classical law held the heir liable provided he was expressly named in the promise, a requirement removed by Justinian (C.8.37(38).13 of AD 530).

According to a principle of classical law no obligation could be made to commence in the person of the heir, the reason being that as his rights and duties were inherited he could not inherit those which had never attached to the deceased (G.3.100).[19] At this time a stipulation in the form 'Do you promise to deliver after my death?' was therefore void. The rule to some extent was relaxed in classical law by allowing a stipulation which could be construed as vesting a right or obligation in a party to the promise while he was still alive even though on the point of death. Thus a stipulation in the form 'Do you promise to pay when I am dying?' was valid, the obligation commencing at the last moment of the promisee's death. A promise to pay me on the day before I die was void. The day could not be determined until after the death and the arrangement was therefore regarded as identical in effect to a stipulation in which the right was to start in the person of the heir.

Justinian abolished the principle that rights or obligations might not begin in the person of the heir, and provided that stipulations to take effect after the death of either party, or using the formula 'before I die', were to be valid in accordance with the intention of the parties (J.3.19.13).

As a general rule the only persons who might acquire rights or duties under a stipulation were the parties to it or their heirs, but the latter only by virtue of the operation of the law of succession. Thus if a person promised that a third party would either benefit or be placed under an obligation, the stipulation was void (J.3.19.3). Hence a promise by A to B that he would pay C gave C no remedy, since he was not a party to the contract; nor indeed did it give B a remedy, even though he was one of the contracting parties, since he had no interest in A's fulfilment of the promise. This particular difficulty could be avoided if A promised a penalty in the event of non-fulfilment (J.3.19.19), the stipulation being in the form, A promises B that he will pay 10 to C and he also promises to pay 10 to B if he fails to pay C. In effect this was a debt to B subject to the negative condition, 'if A fails to pay C'. If, exceptionally, B did have an interest in the performance of a promise in favour of C he could enforce it (J.3.19.20).

One seeming qualification to the rule that a third party could not acquire a right under a stipulation was where a promise was made by a person who was within the authority of another. So, for example, a slave could validly stipulate for a sum of money to be paid to his master who could sue upon the promise. However, this was not conceived as a third party benefiting under a contract, but as an application of the principle that the master acquired what his dependant acquired on the grounds that the dependant was his. The master could not have an obligation imposed on him by the promise of a dependant (J.3.19.4).[20]

Whereas a third party could not receive rights and duties under a stipulation, he could receive payment of what was promised to another so long as the correct form was used, namely, 'Do you promise to give 10 to me or to Seius?' The effect of this was that you could discharge your obligation by paying 10 either to me or Seius, in which case he was liable to me under a contract of mandate (J.3.19.4). However, only I could sue on the stipulation, not Seius.

Another special case involving third persons is considered in J.3.19.4: A promises to pay 10 to B and C. C in this instance is not an alternative person to whom the promisor can discharge his obligation; rather he is directly named as taking a benefit under the stipulation even though he is not a party to it. Such a promise is inherently ambiguous. Is it a promise of 10 each to B and C, or is it 10 divided between them, each taking 5? Albeit all jurists were agreed that the reference to C was void and he therefore could recover nothing, Gaius reports that there was a school dispute concerning the amount that was payable to B. The Sabinians held that he could recover 10 but the Proculians that he could recover 5. Justinian affirmed the Proculian view (J.3.19.4) because of the general presumption of interpretation of an ambiguity in favour of the person placed under the obligation. Thus where the question was whether A should pay 5 or 10, the decision was that he should pay 5.

(f) Guarantors
(J.3.20)

J.3.16 considered the manner in which to provide for a plurality of joint debtors by stipulation. This theme is taken up again in the discussion of guarantors, the sole survivor in the time of Justinian of a variety of stipulatory forms for the provision of personal security collectively known as *adpromissio*. In each a creditor entered into an accessory stipulation whereby he ensured that the same debt owing to him under the main contract, if not recovered from the principal debtor, could be demanded once in full from any one of a number of accessory debtors. The parties stood in a solidary relationship, correal in classical law and simple under Justinian (see J.3.16).

Gaius gives a detailed account of the classical forms (G.3.115 - 127). The earliest joint debtors, dating respectively from the *Twelve Tables* and the 3rd century BC, were called personal sureties or sureties depending on whether the verb *spondere* or *fidepromittere* was used in the ancillary stipulation. The former could be used only by Roman citizens. To the personal surety the creditor put the question, 'Do you solemnly promise to give the same (as in the principal stipulation)?', and to the surety, 'Do you faithfully promise the same?'. The debtor replied in the affirmative using the same verb (G.3.115).[21]

An important restriction applicable to these forms was that they could only be used where the principal debt was contracted by stipulation. Also in both it was only the contracting parties themselves who were bound since liability did not transmit to their heirs. This characteristic is thought to be a reflection of the great age of each institution.

A number of republican statutes were applied to personal sureties and sureties in order to mitigate undesirable features of the correal relationship existing between them. Where there was more than one surety, each was originally liable for the whole of the debt guaranteed and had no right of relief against his co-sureties – nor indeed against the principal debtor until the developed law, when an action on mandate would usually lie against him for reimbursement. A Publilian Act of the 4th century BC, applicable only to personal sureties, provided that if such a person paid the debt, he was to be reimbursed by the principal debtor within six months on pain of paying double damages in the action on expenditure (G.3.127). The Appuleian Act of ca. 200 BC provided that one guarantor who paid more than his share should have an action for the excess proportionately against his co-sureties (G.3.122). The surety sued was nevertheless still obliged to pay the whole debt. Shortly afterwards, the Furian Act, which applied only in Italy, addressed itself to this problem by dividing the obligation into as many shares as there were surviving sureties, each being liable only for his proportion of the whole debt. It further provided that the surety's liability was extinguished two years after the debt fell due (G.3.121). Finally the Cicereian Act of unknown date allowed both sorts of sureties to be discharged from liability unless the principal debtor had given advance public notification of the amount of the debt and the number of personal sureties or sureties he intended to take (G.3.123). The problem was that a surety who had been sued might well have been unaware whether there were any other co-sureties for the debt, and if so, of their number. He might therefore unwittingly have paid out more than his share of the debt as regulated by the Furian Act.

The combined effect of these acts made the provision of security in the above two ways unattractive to creditors, and a more untrammelled form of giving security called 'guarantee' (*fideiussio*) emerged before the end of the Republic. By classical times this was the most commonly used form of personal security.

Guarantee was also contracted by stipulation, the creditor putting the question, 'Do you faithfully authorise the same?', the verb being *fideiubere*. By classical law the giving of security by stipulation using any form other than those appropriate to personal sureties and sureties was deemed to be guarantee (for example, if the question was in the form 'Do you promise the same?' or 'Will you give the same?').

Guarantee had the important advantage that the obligation of the guarantor might be accessory to a principal obligation contracted either by stipulation or in any other way, as by sale, or even a natural obligation (J.3.20.1). The obligation also passed to the guarantor's heirs (J.3.20.2).

The parties to guarantee also stood in a correal relationship in republican and classical law. In time, attempts were made to overcome the same undesirable features associated with correality as were found in surety, but this was accomplished, not by legislation, but by procedural devices. The first of these, appearing in the late Republic, was the *beneficium cedendarum actionum*, whereby a guarantor, if sued by the creditor, could require the latter to cede to him all his rights against the debtor and the other guarantors. The guarantor had to have made his demand before payment and also have been ready fully to discharge the debt.

Later, a written reply of the emperor Hadrian (AD 117-38) allowed any one of a number of guarantors, when sued by the creditor, to require that the action be limited to a proportion of the debt reached by dividing the total among the guarantors solvent at the time of the action (*beneficium divisionis*) (J.3.20.4).

Some legislation dealt both with sureties and guarantors. The Cornelian Act of Sulla (d. 78 BC) limited to twenty thousand sesterces the amount by which a guarantor in any one year could be liable to the same creditor with respect to the same debtor, any excess being void (G.3.124, 125). Also, the Velleian Resolution of the Senate of 46 BC, which is not mentioned by Gaius, prohibited women, *inter alia*, from becoming guarantors except in certain cases (D.16.1). Lastly, although the Cicereian Act did not mention guarantors it was customary for its terms to be observed in this context too (G.3.123).

There were certain other features which sureties and guarantors had in common. At least in developed law, any person who paid a debt had an action on the contract of mandate against the principal debtor for reimbursement (J.3.20.6). Although no guarantor could incur an obligation greater than that of the principal he might incur a lesser obligation. Less meant either that the amount promised was less (5 instead of 10), or that a condition was present which was not attached to the principal contract, or that more time for fulfilment of the accessory obligation was stipulated (J.3.20.5).

Guarantee was the only one of the above forms of personal security to survive under Justinian. In this context there was an important change, again associated with solidary liability. In classical law the relationship existing between the parties to any of the institutions we have discussed was correal. This meant that it was not actual performance in full which extinguished the obligation, but mere joinder of issue with any one of the joint debtors. Consequently a creditor had to take great care in selecting whom to sue. Normally he would be expected to sue the principal debtor first, but should there be reasonable grounds for thinking that he was insolvent, a better course would be to sue one of the accessory debtors instead. Under Justinian the remaining instances of correality were replaced by simple solidarity: it was enacted that the bringing of an action against the main debtor or any guarantor did not of itself extinguish the debt (C.8.40(41).28.2 of AD 531). Only full satisfaction was to have this effect. Consequently Justinian further provided in the *Novels* that in the usual case the creditor must sue the principal debtor first before proceeding against the guarantors (*beneficium ordinis vel excussionis*).[22]

In practice, guarantee would be evidenced by means of a document, the same problems and solutions arising in this context as in the case of the ordinary stipulation. J.3.20.8 reaffirms the rule that a document which stated, or used language permitting the inference, that the requisite formalities had taken place, was to be believed. Probably such a document was only open to challenge within the limits laid down in the pronouncement to the advocates of Caesarea (see J.3.19.12).

IV. Obligations by Writing
(J.3.21)

Strictly speaking this title should deal with obligations created by writing, but in fact Justinian understands by it something different. He does, however, refer to a contract created by written words called 'account entries' (J.3.21 pr). Very little is known about the early history of this contract or indeed its form in classical law, the only substantial account being a somewhat elliptical one in the *Institutes* of Gaius (G.3.128 - 134).[23] The contract was known in the Republic and early Principate, but was probably little used in the time of Gaius and had certainly become obsolete by the end of the classical period.

The head of the family in the Republic kept an account book in which he entered his expenditure and income. The essence of the written contract was the creation of a unilateral obligation of strict law by means of a special entry by the head of the family in this account book. The debtor initiated the contract by requesting the creditor to make a record in the book, the main element of which was that a certain sum of money had been advanced to him: 'Advanced to X 100'. Note that because the entry always alleged the advance of a sum of money, as in *mutuum*, the appropriate remedy was the action of debt. The written entry actually created the obligation to repay, it was not mere evidence of the contract. Gaius distinguishes such 'account entries' from 'cash entries' (G.3.131) in which the obligation arose by conduct and where the function of the writing was purely evidentiary. A cash entry was a record of a *mutuum*, a contract binding by conduct, in which the obligation arose not from the writing but from the handing over of the money. Therefore the account book might contain two different sorts of entry of differing effect, the one being a statement that a loan, albeit fictitious, had been made – 'Advanced to X

100' – and the other being a record of an actual loan. The question is, how were the two entries distinguished?

The answer can be found by looking at the function of the written contract. It was used only for the purpose of novation, which it might do in one of two ways (G.3.128):

(1) By transfer from a transaction to a person. In this case the parties and the debt remained the same but the ground of the debt was transformed. For example, the head of the family may have had a number of dealings with person X, who was indebted to him under a number of different transactions, say, *mutuum*, stipulation, and sale. The creditor might add up all the different amounts and write in his account book that this sum had been advanced to X. The effect was to extinguish all the previous obligations, which were replaced by the obligation to pay under this contract. The advantage was that the creditor need now only bring a single action to recover the whole sum due instead of the separate actions on *mutuum*, stipulation, and sale.

(2) By transfer from one person to another. In this case the creditor who was owed money by A wrote, at the request of A and B, that what was owed by A had been advanced to B. A's debt was cancelled and replaced by B's obligation to pay.

The latter form suggests that if a statement appeared in the account book merely to the effect that a certain sum of money had been advanced, no obligation was created by the writing. In fact, as is shown by the plural 'account entries', there must have been two separate entries, one which recorded the advance of the sum of money and the other which, in the case of transfer from a transaction to a person, recorded the receipt of what had been due under the earlier transactions or, in the case of transfer from one person to another, recorded the receipt of the money from the first debtor.[24] The consent of the debtor was always required, but it is not clear how this was indicated. The contract could not be subject to a condition.[25]

As we have said, the written contract became obsolete during the classical period. However, documents by which a debt was acknowledged, 'Advanced to X 100' or 'Advanced to X 100 as a loan', were still prevalent. Indeed, increasing recourse was had to the device of a fictitious loan, which was valuable as a means of outflanking the legal restrictions on the amount of interest chargeable. A creditor lends 50 but obtains from the debtor a written acknowledgement that 100 has been received, the interest in effect being 100%. Alternatively, having executed such a document, a dishonest creditor might fail to hand over any money at all. Although documents in this form were mere evidence of a *mutuum* and there was therefore no obligation in law unless the money was in fact handed over (and even then it was only to repay the actual amount delivered), in practice the document could readily appear to have a dispositive effect. A creditor sues on the contract of *mutuum* and produces the debtor's written acknowledgement as proof of the loan. The only defence available to the debtor for most of the classical period was the defence of fraud, under which he had the onus of proving that the amount stated in the document had not been received. This would normally prove extremely difficult since the creditor held the document signed by him which stated that the money had been received. To deal with this problem, within the classical period but perhaps towards its end, a special defence, whose distinctive feature was its negative formulation, was allowed for cases in which an action was brought on a *mutuum* evidenced by a written acknowledgment. In response to the creditor's action claiming repayment of the spurious loan, the debtor was allowed a defence that the money had not been paid. The effect was to shift the burden of proof on to the creditor who, independently of the document, had to prove the opposite of this negative statement, namely that the money had been advanced by him. In the first instance the defence was available only for one year from the date of the alleged loan. Since this allowed the creditor simply to wait until the period had expired before suing, the law further buttressed the position of the debtor by allowing him within the year to sue for the return of the document by the action of debt. In such an action he would succeed unless the creditor could prove that the money had been paid, or that there was an actual debt of some other kind (C.4.30.5 of AD 223). Diocletian extended the time period for the availability of this remedy to five years, but it was reduced by Justinian to two.[26]

In his title on obligations created by writing (J.3.21), Justinian deals with documents by which a debt is acknowledged. He says that provided no stipulation has been made, such documents can be said to create an obligation because after the lapse of the period of two years within which the defence of money not paid was available, the debtor was bound by their terms. However, this is incorrect. These documents never created the obligation, they were only ever evidence of it. But after the two years they were evidence which would have been extremely difficult to overcome. For this reason Justinian says that the debtor incurs an obligation by writing, but nevertheless what he is doing is treating a matter of evidence as one of substantive law.

V. Obligations by Agreement
(J.3.22 - 26)

The principle of liability in the contracts comprising this group is that obligation arises from agreement alone. The principle did not mean that any agreement was legally enforceable. It was restricted to certain standard cases, namely the commercially or practically most important transactions of sale, hire, partnership, and mandate. These developed in the last two centuries of the Republic and so, with the exception of some of the contracts binding by conduct, were the last of the nominate contracts to emerge. They were flexible legal instruments characterised by bilaterality and good faith.

(a) Sale
(J.3.23)

Gaius' account of sale is surprisingly sparse; it confines itself to the rules governing formation of the contract, in particular those on price. It says nothing of the rights and duties created by the contract or of the effect of conditions or special terms. Justinian's account is slightly fuller, in that it includes sections which set out the distinction between conditional and unconditional sales and it summarises the rules on risk.

Gaius states that the contract of sale is concluded as soon as there is agreement on the price, even if no down-payment (*arra*) is made (G.3.139). He adds that if a down-payment is made, it is to be treated merely as evidence that the parties have passed from the stage of negotiations and have resolved upon a binding contract. J.3.23 pr follows Gaius in the first part of his treatment, but then qualifies the Gaian position by stating that these rules apply only to 'unwritten sales', no changes having been made to such sales. With respect to sales to be concluded 'in writing', a pronouncement issued by Justinian had made validity conditional upon the completion of certain formalities (C.4.21.17 of AD 528). Contracts which the parties had agreed to put into writing were not to have legal effect unless (1) they had been written down in final form and signed by the parties, or (2) they had been drawn up by a notary and delivered to the parties.[27] The pronouncement further provides that whether contracts of sale are to be completed in writing or not, if a down-payment has been made, in the absence of express agreement concerning its destiny, the buyer is to forfeit the down-payment if he repudiates the contract, and the seller is to restore double if he repudiates.

Justinian's provisions concerning down-payment cause some difficulty in the interpretation of his account in the *Institutes*. He purports to have made no change in the law relating to unwritten contracts of sale, where, if made, the down-payment was purely evidentiary, yet at the same time he applies the rules giving the down-payment the function of a penalty to both unwritten and written contracts. A number of attempts to solve the problem have been made.[28] The most straightforward is that which assumes that in classical law and throughout the post-classical period, the down-payment was commonly treated as providing the unit for the measure of damages awarded against a party who simply repudiated the contract of sale: forfeiture by the buyer and payment of double by the seller. Thus Justinian, in applying his penal rule to unwritten as well as to written sales, was innovating only in the sense that he was stating as a rule of law what, in respect of unwritten sales, had been a matter of common practice.[29]

For both Gaius and Justinian the most important rule for the conclusion of a valid contract was agreement on the price. They say nothing about the further requirement that there be an existing object which the parties had agreed should be bought and sold. The general rule is that a sale was void if the object to be sold had never existed or had ceased to exist prior to the conclusion of the contract (Paul, D.18.1.15 pr; D.h.t.57; Papinian, D.h.t.58).[30] There must be an identifiable object in existence at the time of the making of the contract. Two apparent exceptions were allowed, the sale of future crops and of offspring (Pomponius, D.18.1.8 pr); these were usually dealt with as typical, but not invariable, examples of classes of sales known as the 'sale of an expected thing', and the 'sale of a chance' (Pomponius, D.18.1.8.1).

In the first case a contract was concluded for the sale of an object not yet in existence but which was expected to appear at a future date. The sale was conditional upon the object to be sold coming into existence, and only in that event was the price payable. However, the seller was liable under the contract if he prevented the object from materialising. In the second case the contract was for something possibly in existence at the time of agreement but which might very well not materialise as the object of that particular sale, such as 'whatever should be caught in the next haul of the fisherman's net'. Such a contract was immediately valid and the buyer was under an obligation to pay the price whether anything was caught or not. The classical jurists came to speak of it as the sale of a chance.

Some scholars distinguish these two classes of sale, not by reference to the differing object of the contract, an expected thing or a chance, but by reference to the price, which they argue was fixed at so much per unit in the first class but at a lump sum in the second. One result of this approach can be seen in respect of sales of future crops. They are classed as 'sale of an expected thing' or 'sale of a chance' depending on whether the price is per unit or a lump sum.[31] Yet there is no strong evidence to suggest that these two general classes of sale were distinguishable on the basis of price; nor indeed does the main *Digest* text (Pomponius, D.18.1.8) speak of anything called 'sale of an expected thing'; it refers only to the sale of future crops or offspring, which it identifies as conditional. The implication is that all sales of future crops and offspring (where the parties have every anticipation that the object will appear) were regarded as conditional on the object materialising whether the price was fixed at so much per unit or, as it certainly could be, at a lump sum.

A point obliquely made by Justinian, though absent in Gaius, is that the object must be capable of ownership by a private individual (J.3.23.5). Certain classes of object could not be bought or sold, in particular free persons and sacred or religious ground. A separate issue was the extent to which a contract for their sale was enforceable as between the parties. In order to make Justinian's treatment fully understandable it is necessary to give a statement of the classical background, which itself is a matter of some controversy. In early law, sales of all the above-mentioned objects were void, giving the buyer an action of debt for the return of the price or, if the seller had been fraudulent, a claim for damages in the action for fraud. However, in classical law the history of sales of free men and sacred or religious ground was different. The sale of a freeman came to be regarded as valid, first when both buyer and seller were unaware of his true status, and later even when the seller was fraudulent (Licinnius Rufinus, D.18.1.70). The motivation for the recognition of the validity of the sale was a desire to furnish the buyer with a remedy whereby he could claim damages. This desire was first felt when the seller was innocent, because in that instance the buyer had only had the action of debt for the return of the money given as price. On the other hand, sales of sacred and religious land were always void during the classical period (Ulpian, D.18.1.22; Paul, D.h.t.23). However, notwithstanding the invalidity of the contract (and surprising from the point of view of principle), as a special remedy an action on purchase came to be given to the buyer, but in this instance only if the seller was fraudulent (Modestinus, D.18.1.62.1).

J.3.23.5 is explicable by the fact that, at the hands of Justinian, there had been cross-fertilisation between these two distinct fields of law. The result is unsatisfactory because of the methods used by the compilers in constructing the *Institutes*. Firstly, as is shown by an interpolated text (Pomponius, D.18.1.4), Justinian extended the validity of the sale in the case of the freeman to the case of sacred and religious land, so long as the buyer was in good faith. Yet, as is clear from J.3.23.5 itself, the circumstances in which an action on purchase was given to the buyer of sacred and religious land in classical law are applied to the case of the freeman; that is, the action is now also given in this case only when the seller was fraudulent. But it is highly unlikely that the compilers intended to restrict the availability of the purchase action in this manner. They intended that sales of all these objects should now be treated on the same basis. In expressing this in the *Institutes* they have utilised a classical text concerning sacred and religious ground from which, having added the case of the freeman, they failed to remove the limitation of the contractual action to the case where the seller was fraudulent. The limitation existed in classical law in this context only because the action on purchase had been given as a special remedy for fraud in a void sale. Once the validity of the sale had been recognised, it would have made no sense to allow the action only against the fraudulent seller.[32]

Sale of what already belonged to the buyer was void (Pomponius, D.18.1.16; cf. Paul, 18.1.34.4) but, because the seller was obliged only to guarantee peaceful possession of the object (Ulpian, D.18.1.25), sale of what belonged to a third party was valid unless both parties knew the object was stolen or improperly obtained (Ulpian, D.18.1.28; Paul, 18.1.34.3).[33]

Both Gaius and Justinian take agreement on the money price as the decisive moment for the conclusion of the contract. One reason for this emphasis is that it was a money price given by the buyer which ultimately was held to distinguish sale from an agreement to exchange goods. This position was not reached without controversy (J.3.23.2). The Sabinians, relying upon the archaic antecedents of sale, had seen no reason to distinguish a transaction in which money was given for goods from one in which goods were given for goods. The Proculians, however, held that under a contract of sale it was necessary to distinguish which party was buyer and which seller, since the rights and duties of each were different. This distinction could not readily be made in exchange. Even if it could be made, as Sabinus suggests (G.3.141), by treating the person who offered his goods for sale as the seller, the result would have been unfair since the 'buyer' was obliged to transfer ownership of the object he gave as price but the 'seller' need only guarantee vacant possession of the object he gave. By the end of the classical period the controversy had been decided in favour of the Proculians that the price had to be in money (Paul, D.19.4.1).

An arrangement in which the price was partly in money and partly another object is likely to have been regarded as sale by the early Sabinians. It also meets the major objection of the Proculians to exchange being treated as sale, since the buyer will have been identifiable as the person who gave the part-money price. Nevertheless, there is little evidence to suggest that such a doctrine gained independent support. There are some cases of sale for a money price plus an undertaking by the buyer; for example, I purchase half an estate for a money price and further provide that I shall rent the remaining half for a specified period (Javolenus, D.18.1.79). In developed classical law such ancillary terms were enforceable by the seller's action on purchase, which might suggest that they were treated as part of the price. However, Daube shows that they were construed as imposing a subsidiary duty on the buyer separate from his principal duty, which was to pay the money price.[34] Therefore when Justinian says that the price must be in money, this should be taken as an unequivocal statement that it must be exclusively in money.

Both Gaius (G.3.140) and J.3.23.1 state that the price must be certain; '100 pounds', for example. Sale at a 'reasonable' price was void. But a practical difficulty arose where buyer and seller wished to conclude a contract but could not agree on the exact amount of the price. One solution was to leave it to be fixed by an independent named third person. Did such an arrangement satisfy the rule requiring certainty? Although no definite sum had been agreed, a definite method had been established for fixing the price. Gaius simply reports a difference of opinion among the jurists: Labeo (a Proculian) and Cassius (a Sabinian) held that the arrangement was void, whereas Ofilius and Proculus held it to be a sale. Justinian treated it as a conditional sale; it was valid but subject to the condition that the named third party fixed a certain price. If he did not, the sale was void.

It is not clear whether Justinian was innovating in this respect or whether he was adopting a solution which had been aired in the classical period.

Sale at a price to be fixed by the buyer was probably void in classical law but, on the analogy of price to be fixed by a named third person, was treated as conditional on the buyer setting a price in the time of Justinian (Gaius, D.18.1.35.1).[35] Sale for 'the amount of money in my safe' was valid. The price was certain albeit the exact amount may not have been known to the parties (Ulpian, D.18.1.7.1). Such a sale was again deemed to be conditional but, in contrast to 'sale of a chance', was valid only if there was in fact some money in the safe; for otherwise the agreement raised the possibility of sale without a price, which was void.[36]

Two further requirements in respect of price are mentioned neither by Gaius nor by Justinian. They were that the price should be real and just. The first, which is classical, is designed to prevent what is in effect a gift masquerading as a sale; thus a purported sale with a derisory price is treated as a sale which lacks a price and therefore void, though the transaction might take effect as a gift. The same applies if what appears to be a genuine price is stated yet it can be shown that the seller had no intention of exacting it. Where, however, the seller, wishing to benefit the buyer, sells for less than the true value, the sale is valid unless it is between husband and wife (Ulpian, D.18.1.38). As regards the requirement that the price be just, the later law, perhaps even as late as Justinian, restricted the extent to which advantage might be taken of a seller by allowing the latter to rescind where the contract fixed a price at less than half the value of the land sold, unless the buyer chose to make up the difference between the price agreed and the just price.[37]

Neither of the institutional works considers the effect of mistake. The general principle is that if the minds of the parties do not meet on the fundamental questions of price and object there is no sale because of lack of agreement. With respect to price the exception was allowed that where the seller understood the price to be lower than the buyer did, the contract took effect at that price (see Pomponius, D.19.2.52). With respect to the object, the rule that came to be settled was that there was no contract if 'one thing was sold for another'. There was no difficulty in applying this rule in cases where the seller thought, for example, that he was selling one distinct plot of land or slave and the buyer that he was buying another (Ulpian, D.18.1.9 pr).[38] But there was uncertainty in the case where both had agreed on the same physical object but had differing conceptions as to its nature or qualities.

The jurists in the mid-classical period held that there was a valid sale where the parties had agreed on the physical object even though they were mistaken as to its nature (Ulpian, D.18.1.9.2). This was no doubt the majority view up to the time of Marcellus. Subsequently the jurists modified their attitude and introduced what has come to be called the doctrine of *error in substantia*. If there was a mistake as to the essential nature or substance of an object the contract was void; if there was a mistake merely as to a quality which it possessed the contract was valid.

Thus where vinegar was sold as wine, or gold or lead was sold as silver, the mistake was treated as one of substance and the contract held void. On the other hand a mistake as to the quality of a metal, the belief that it was pure instead of an alloy, did not avoid the contract. Equally, the contract was void where there was a mistake as to the sex of a slave but not where the mistake was as to virginity.

While the rule avoiding a contract certainly applied in cases in which the seller innocently represented the object, which was not present or could not be seen by the buyer, as possessing a nature or character which it lacked, it was not limited to such cases (Ulpian, D.18.1.11).[39] Similarly, though the jurists primarily envisaged mistake by the buyer, the rule also applied if there was proof of mistake on the part of both parties (Ulpian, D.18.1.14).

If the contract was void on the ground of mistake, the goods, if received, were to be returned to the seller, and the buyer had an action of debt to recover the price paid. If the seller was dishonest, in theory the buyer had the action for fraud to recover what he had lost.[40] If the mistake was not such as to avoid the contract, in classical law the buyer was required to keep the goods and pay the price, but might have a remedy on the ground of defect in the goods sold. In the law of Justinian he might choose between rescinding the contract or affirming it and suing for damages on the ground of the defect.

Sale might be contracted absolutely or subject to a condition (J.3.23.4). Conditions made the full operation and the existence of the contract depend on the occurrence of some future uncertain event. Although the contract was concluded, with the result that neither party could withdraw unilaterally, it was imperfect and the obligations of the parties were not enforceable until the condition was fulfilled. Thus, in the example given by Justinian, the sale of Stichus will not be fully operational until the buyer has approved the slave. If he does not do so within the time agreed the sale is void. A conditional sale is to be distinguished from that which is immediately fully operative but which contains a term dissolving the sale under certain circumstances: for example, where the parties defer payment of the price but provide that the seller may annul the contract if it is not paid by the agreed date. In such cases it was not the contract itself which was regarded as conditional but its dissolution (Ulpian, D.18.1.3).

Once the contract had been concluded, risk with respect to accidental loss or damage passed immediately to the buyer even if the thing had not yet been delivered (J.3.23.3). But Justinian's discussion of risk is predicated upon the notion of fault; this is interrelated with, but distinct from, the concept of risk. The buyer bears the risk in the sense that he is still bound to pay the full price if the goods are lost or damaged independently of his or the seller's fault. Where he had assumed an insurance liability (*custodia*) with respect to the goods sold (J.3.23.3a) the risk was on the seller since, without fault on his part, he bore all loss except that caused by an act of God. If it is the result of the seller's fault, whether deliberate or not, that the goods are lost or damaged before delivery, he bears the loss but not the risk. In this case it is on the ground of his fault that he is liable to the buyer, who may claim a set-off against the price.

Justinian's account of risk is oversimplified and in one respect misleading. He treats total destruction of the goods as no different from damage to them. In this context, from the sale that is immediately perfect and to which the above rules on risk applied at once, we have to distinguish sales that, though concluded, are imperfect (Paul, D.18.6.8 pr; Gaius, D.18.1.35.4 - 7). A sale is imperfect in three typical cases: where it is conditional, until the condition is fulfilled; where the object is undetermined, having to be taken from stock or chosen from alternatives; and where the price still has to be fully ascertained, as where I buy the contents of your grain store at so much per unit. If the goods were totally destroyed or lost before the contract became perfect, it was treated as void for lack of an object and the buyer was released from his obligation to pay the price. If the goods were merely damaged, once the contract became perfect, the risk retrospectively passed to the buyer. In effect, in an imperfect sale risk of total destruction fell on the seller, of deterioration on the buyer.

It is sometimes thought that what Justinian states to be the result of a special agreement as to insurance liability (J.3.23.3a) was the normal rule in classical law. In all cases, it is thought, the seller was liable for the safekeeping of the object and therefore the buyer bore the risk only for such loss or damage as occurred through an act of God. The seller was liable for such loss or damage as occurred merely through accident, even though he had in no way been at fault. However, the texts are not clear and the matter must be regarded as doubtful.[41]

The two most important obligations incumbent upon the seller are mentioned by neither Gaius nor Justinian, namely, his obligation to make good a loss caused through defects in the goods sold, or a loss caused through failure to ensure that the buyer acquired peaceful possession. One possible reconstruction of the history of the rules on the seller's obligation with respect to defects in goods sold is as follows.

Taking the *Twelve Tables* as a starting point, in the absence of express undertakings by stipulation and of a restricted liability in respect of land (arising, not from the contract of sale, but the mancipation),[42] the seller was not liable for defects in goods sold. In the late Republic the jurists introduced a liability under the contract of sale itself for non-disclosure by the seller of certain classes of defect known to him. The non-disclosure was regarded as fraud and a breach of the good-faith character of the contract.[43] At first the liability was limited to legal defects in land, such as the existence of a servitude, but it was later extended to material defects, at least in moveables (Ulpian, D.21.1.4.4). There does not appear to have been a general liability under the action on purchase for non-disclosure of defects of which the seller was unaware. However, after the time of Labeo (late Republic and early Empire), the seller may have been held strictly liable if the goods possessed a fundamental defect which made them totally unsuitable for their normal purpose, as where a vase turned out to have a hole in it (Pomponius, D.19.1.6.4).[44]

In addition to his responsibility for fraudulent non-disclosure, the seller was liable under the contract if he represented the goods as possessing qualities which they lacked, though he had himself believed the representation to be true. By 'defect' in all cases is meant a defect which is latent, one not obvious to the buyer on reasonable inspection. The measure of damages was the difference between the price paid and the value of the defective goods, unless there was fraud, in which case the action on purchase lay to the extent of the buyer's interest, damages including consequential loss.

In the Republic the edict of the aediles introduced better protection for the buyer with respect to the sale of slaves and cattle in the market (Ulpian, D.21.1.1.1; D.h.t.38 pr). In these cases the seller was made strictly liable where the slave or animal had a disease or material defect, even where he had not known about it and had made no misrepresentations. The edict required the seller to state, in the case of a slave, whether he had any disease or other physical defect or whether he was a runaway or wanderer or had committed a delict and, in the case of cattle, whether the animal had a disease or physical defect. Failure to make the statement gave the buyer a choice between rescission of the contract within two months (*actio redhibitoria*) or damages within six months (*actio quanti minoris*) even if no fault had in fact materialised (Gaius, D.21.1.28). Whether or not the statement was made, the buyer had a longer period in which to rescind (six months) or sue for damages (one year) where the slave or animal turned out to have a disease or material physical defect. In addition, the seller was made liable under the edict for any statement he made affirming the presence of some quality or the absence of some defect not covered by the other rules, the buyer again having a choice between rescission or damages.

The importance of the edict lay primarily in two factors: the strictness of the liability imposed on the seller with respect to certain categories of defect, making him liable for innocent non-disclosure, and the introduction of the remedy of rescission as an option available to the buyer.[45] Even in classical law there may have been some extension of the provisions of the aedile's edict to sales other than those of slaves or cattle in the market. Certainly by the time of Justinian it had become settled that the edictal rules applied to all sales, irrespective of subject matter, whether or not concluded in the market. The result was to introduce for the law of Justinian what may be termed an implied warranty in the contract with respect to defects in the goods sold. The seller was strictly liable whether or not he knew of the defect or made any misrepresentation. In principle the buyer now had the action on purchase for rescission of the contract or for damages, including consequential loss if there was fraud. Possibly practice determined that rescission was granted only for serious defects.

The seller was not bound to make the buyer owner of the property sold, but should he deliver what he knew he did not own he was liable to the buyer in the contractual action on the ground of fraud for any loss suffered by the buyer (Proculus, D.18.1.68.2). He was, however, bound to put the buyer into undisturbed possession. The development of the rules relating to liability on the ground of eviction occurred through the use and interpretation of two traditional stipulations enforceable by the action of debt. By the *stipulatio duplae* the seller promised to pay to the buyer double the price should a third person be able to prove a better legal title and secure the buyer's eviction (Pomponius, D.21.2.16.1). It was normally given for sales of slaves or land but was not confined to such cases. By the *stipulatio habere licere* in its original form, the seller promised that neither he nor his heirs would subsequently perform any act by which the buyer's possession would be disturbed. It seems that at least some jurists by the end of the classical period were prepared to construe the *stipulatio habere licere* as applicable to the case in which the buyer's possession was disturbed by a third person, provided the stipulation was not expressly limited to acts of the seller or his heirs (Ulpian, D.19.1.11.18).[46] Probably even before the end of the classical period the stipulations were implied, where relevant, in the contract of sale, and the buyer had the action on purchase for the appropriate damages. In the case of land (unless regional practice was otherwise), slaves, and valuables, the buyer in the event of eviction might sue as though the *stipulatio duplae* had been given, and in the event of other things, as though the *stipulatio habere licere* had been given (cf. Ulpian, D.21.2.37 pr, 1; Gaius, D.h.t.6).[47]

(b) Hire
(J.3.24)

Although it is generally thought not to be discernible in the Roman sources,[48] modern commentators follow the humanists' classification of the contract of hire (*locatio conductio*) as embracing three kinds of transaction: (1) the hire of property (*locatio conductio rei*), in respect of which no distinction was made between land and moveables; (2) the placing out of a job of work to be done (*locatio conductio operis*); here the contract was for one specific, clearly defined task which was usually of a physical nature such as the agreement with a goldsmith to make rings out of gold which I supply, or with a builder to erect a house on my land, and even the agreement to train an apprentice (Julian, in Ulpian, D.19.2.13.4); and (3) the hiring out of services (*locatio conductio operarum*); the contract applied only to those of low social status and hence in practice governed services which were regarded as menial, as where a man hired himself out as a gardener or a labourer. Thus, the contract was hire wherever one party, the lessor, agreed to place at the disposal of another, the hirer, some object, a piece of work to be done or his services. This must be done in return for a fixed sum of money called the charge which, in the case of hire of property or services, was paid by the hirer but, where a job of work was placed out, by the lessor.

Justinian says that hire is controlled by the same rules as sale, Gaius, more accurately, that it is formed under similar rules. They mean that hire, like sale, is concluded by agreement on the object of the contract and on the charge. Neither institutional treatment defines the object of the contract; both concentrate almost exclusively upon the criteria by which a contract was deemed to be hire rather than some other contract. The essential criterion was that of charge (*merces*) which, as with price (*pretium*) in sale, must be fixed and certain. If services were to be rendered on the understanding that no return would be made, the contract was mandate. If there was to be a return but either it was not to be in money, as where we agree to exchange things for temporary use, or its quantity was uncertain, as where the charge was to be fixed by a named third party, there was a problem. In both these instances Gaius leaves open the question whether the contract is hire (G.3.143, 144). In the latter case Justinian distinguishes according to the facts. If the charge is to be fixed by a third person nominated in the initial agreement, the same rule applies as in sale. The contract is valid but conditional upon the third person fixing the reward. If the agreement is merely that a reward will be paid, but no mechanism for fixing the charge has been established, the contract is not hire but an innominate contract enforceable by the action with a special preface (J.3.24.1). Justinian also says that this is the appropriate remedy where we have agreed to exchange things for temporary use. However, whereas in sale the price had to be exclusively in money, this was not always the case in respect of the charge. It was permissible in the case of agricultural tenancies for the charge to be a fixed proportion of the crops.

Where the amount of the money charge had been settled at the time of agreement, there were sometimes problems of demarcation between hire and sale. Gaius singles out three different cases (G.3.145 - 147): where something is let in perpetuity, where gladiators are hired out upon terms that so much is to be paid for any who comes off unharmed in the forthcoming contest but so much (a greater amount) is payable for any who is killed or injured, and where a goldsmith is commissioned to make rings for a fixed amount.

In Gaius' time the problem of property let in perpetuity was confined to land owned by municipalities let upon terms that it would never be taken away from the tenant and his heirs so long as the rent was paid. Such an arrangement differed from a normal case of sale since, although the tenant was able to dispose of the property as if he were the owner, his right to the land was dependent on his continuing to pay the charge; yet neither was it a normal contract of hire under which the tenant had no right to possess in perpetuity. Nevertheless Gaius states that the majority opinion treated the arrangement as a contract of hire. Clearly some doubts remained, probably accelerated during the post-classical period when the emperor and private individuals also adopted the practice of making grants of land in perpetuity in return for an annual rent, the grant to be forfeited on failure to pay. Eventually the emperor Zeno (AD 474-91) enacted that this was neither hire nor sale but a distinct type of contract to be termed *emphyteusis* (see Chapter 3, section IV(e) above). He also settled the incidence of risk as recorded in J.3.24.3.

The case of the gladiators put by Gaius is interesting because it shows that an agreement might be subject to a condition which determined the nature of the contract itself. Gaius states that there was deemed to be a conditional sale or conditional hire in respect of each gladiator, the outcome determining which contract was operative for that individual. A difficulty concerns the remedy, if any, which was available before the fulfilment of the condition. The case is not mentioned by Justinian since by his time gladiatorial contests had been banned.[49]

Gaius' discussion of the case of the goldsmith is followed almost verbatim by Justinian (J.3.24.4). The distinguishing factor was that in sale there was a transfer of ownership to Titius of the materials supplied by the goldsmith. One important practical exception to this (probably Sabinus' solution to the controversy) was the case in which A agreed to furnish the materials and build a house for B on B's land. In this instance the contract was hire, not sale, even though A supplied and thus transferred ownership in the materials to B. The reason was probably that the materials acceded to the land and hence were owned by B prior to the completion of the contract (cf. Sabinus, in Pomponius, D.18.1.20; Paul, D.19.2.22.2).

In all cases, the fundamental incidents of the contract were implied by good faith: they need not be expressly stated and might be varied by agreement. However, in considering the question of some of the rights and duties imposed on the parties, altogether omitted by Gaius and barely noted by Justinian, it is best to treat separately the three kinds of hire. The tenant of land, in common with other hirers, acquired mere physical control, not possession, of the property and hence, if ejected either by a third person or by the landlord himself, had no possessory remedy, merely a possible contractual action for damages. He was liable to the landlord for damage to the property sustained through his deliberate or non-deliberate fault, the latter being measured generally by the standard of care expected of a careful owner. On the other hand the landlord was required to keep the buildings in proper repair; that is to say, he sustained the burden of normal wear and tear or accidental damage, though, in the case of farms, the tenant was under an obligation to keep the land in proper cultivation.

Risk of accidental destruction or serious damage was on the landlord in the sense that he was not entitled to claim the charge to the extent that the tenant had been unable, through the accident, to have proper enjoyment and use of the property. Thus a failure of crops due to adverse seasonal conditions entitled the tenant to a rebate on the charge, to be made good in later more prosperous years (Papinian, in Ulpian, D.19.2.15.4).

Contracts for the hire of services call for no special comment, the general point being that both parties were liable with respect to loss due to fault. The same principle applies to contracts for a job of work to be done. But here there were important rules regarding risk. If the work was destroyed by an act of God, such as an earthquake, the party placing the order bore the loss in the sense that he still had to pay the agreed charge. The same rule was applied where the work was destroyed or impaired through some defect in the materials which he supplied. However the risk of mere accidental loss not caused by an act of God was to be borne by the party carrying out the work. Once the work had been completed and approved, all risk was on the party who had ordered it.

An important special rule applied to contracts for the carriage of goods by sea. Implied in the contract was a term derived from the maritime practice of the city of Rhodes that should goods have to be jettisoned to prevent shipwreck, in the event of the ship being saved, the loss of the jettisoned goods was to be shared between all those conveying goods and the shipmaster himself in proportion to the value of the goods preserved and the ship itself (Paul, D.14.2.2). The owner of the jettisoned goods might sue the shipmaster for their value, less that proportion which he had to bear as loss himself, leaving it to the shipmaster to recover proportionately from the other persons who had consigned goods on board.

(c) Partnership
(J.3.25)

Buckland defines partnership as 'the union of funds, skill, or labour, or a combination of them, for a common purpose which often had, but need not have, profit for its aim'.[50] It was their common purpose, which must be neither immoral nor illegal, which distinguished partners from mere common owners, albeit that partners were often common owners of the partnership property.

The partnership known to Gaius and Justinian was a consensual contract in which the essential element for the conclusion and maintenance of the contract was the consent of the individual partners (D.17.2). However, prior to the development of the consensual contract, there was a form of partnership peculiar to Roman citizens called 'ownership undivided' which was constituted by operation of law. In origin this was a family arrangement designed to facilitate the management of family property.

Gaius, the only source (G.3.154a), states that on the death of a head of the family his immediate heirs – he also calls the arrangement a partnership of brothers (G.3.154b) but in principle immediate heirs would have included unmarried daughters and a wife in marital subordination – constituted a partnership in the sense that all were co-owners of the property left by the deceased. An arrangement similar to 'ownership undivided' could also be established by persons other than immediate heirs through the completion of a set action in the law before the praetor (G.3.154b). Although it was obsolete in the classical law, this institution nevertheless had some influence upon the development of the rules governing the consensual contract.

A peculiarity of 'ownership undivided' was that, although the partners were co-owners, any one of them, without authorisation, could perform legal acts, binding on them all, with respect to the property held in common. Gaius mentions the manumission of a slave and the mancipation of items of property, but does not make clear whether these cases were merely examples or exhaustive of the circumstances in which one person could bind his partners.

Various kinds of partnership, derived from agreement, might be established in the developed law depending on the range of the common purpose (Ulpian, D.17.2.5). (1) Where the intention was to pool all property and hold it in common, the partnership was 'of all worldly wealth'. This was not really a business arrangement but was used mostly within the family as a replacement for the old 'ownership undivided'. Depending on the intentions of the parties the other forms of partnership were: (2) for one particular task, for example where neighbours buy a piece of ground to prevent obstruction of their light (Mela, in Ulpian, D.17.2.52.13); (3) for a single line of business such as the buying or selling of slaves; or (4) for all business affairs. Where the nature of the partnership was unclear the presumption was that it was of this last sort (Ulpian, D.17.2.7). Gaius (G.3.148) and Justinian (J.3.25 pr) mention only type (1) and, no doubt the most common, type (3).

All the consensual forms of partnership had certain important rules in common. Since the legal consequences of the partnership were derived fundamentally from the consent of the partners, the legal position of each individual only changed with respect to that of the other individuals composing the partnership. Formation of the partnership affected the rights and duties of the individual partners amongst themselves, not those of third persons. Thus, alienation of partnership property by one partner could convey no more than his own share. Similarly, except in the case of banking partnerships, a third person dealing with a partner dealt with him (from the legal perspective) as an individual and not as representative of a partnership. For example, if I conclude a sale with partner X it is only against or by X that I can proceed or be sued. X's partners were obliged to make their contribution to expenses he incurred in the course of the business and, in the event of their refusing without good cause, his ultimate course of action was to proceed against them on the contract of partnership.

Especially in the commercial or business context, the fact that a partner could be dealt with only as an individual proved inconvenient, and ways around the restriction were developed by the law. Most importantly, where the partners had given a mandate to one of their number to conclude a transaction with a third person, the late classical jurist Papinian gave the third party an action 'as if institorian' against any of the mandators (D.14.3.19 pr).[51]

The partners had to make their agreed contribution to the venture, otherwise it might be indistinguishable from reciprocal gift, and they had to take their share in profit and loss. Generally speaking, their relations were governed by the principle of good faith but, deriving from its origin in the old partnership of 'ownership undivided', a distinctive feature of this contract was that good faith was interpreted in the light of the 'brotherhood' existing between partners. The effect of this notion was to impose on the parties special rights and duties consistent with those existing between persons who stood in such a close relationship. By contrast with, for example, buyer and seller, who had different rights and duties enforceable by different remedies under the contract of sale, partners had the same rights and duties in respect of each other enforceable by the single good-faith action on partnership. The assumption was made that the relations of brothers should be governed by the principle of equality. This idea is clearly reflected in the discussion in J.3.25.1 - 3 of distribution of profit and loss. In the old 'ownership undivided' profits and losses were shared equally by the members. No doubt this was often the position under the consensual contract but, precisely because it was founded on consent, the question naturally arose, could the partners validly agree to an unequal distribution of profit and loss? There seems to have been no difficulty in admitting that one partner might take, say, two-thirds of profits if he was also to take the same proportion of losses. But what if he were to take a greater share in profit than loss or vice versa? The two late republican jurists Q. Mucius Scaevola and Servius Sulpicius differed on the point. The former maintained that such an arrangement was contrary to the nature of partnership, but the latter successfully argued that the parties might vary the distribution of profit and loss even to the point where one partner could take a share of profit but bear no loss at all. However, it came to be settled that any arrangement under which a partner was excluded altogether from profits and bore only losses was void. Also, in the absence of express provision for profit and loss it was held that the shares in each were to be equal irrespective of the size of each partner's contribution. A number of other rules reflect the particularly close relations of partners; for example, condemnation in the action on partnership was limited to what the individual could pay (Sabinus, in Ulpian, D.17.2.63 pr) and, most remarkably, merely to bring the action at all resulted in dissolution of the partnership on the grounds that this was inconsistent with brotherhood (Proculus, in Paul, D.17.2.65 pr).[52]

Justinian's formulation of rules on the standard of liability required of partners in their dealings with each other suggests that originally liability was only for malice (J.3.25.9). He compares the contract of partnership with that of deposit, the point being that just as a depositor chooses his depositee and so must take the risk of the latter's carelessness, so too partners must be deemed to know the extent to which each can be relied upon. The idea of liability having been restricted to malice is supported by the fact that a person condemned in the action on partnership incurred infamy. However, classical jurists make it clear that partners were liable to each other with respect to non-deliberate fault and that this was judged, not by reference to the very highest standard of care, but to the same standard as the partner usually displayed in his own affairs (Celsus, in Ulpian, D.17.2.52.2; Gaius, D.17.2.72). There is some controversy as to whether the limitation of liability to such carelessness was classical or post-classical.

Dissolution of the partnership is dealt with at length by Justinian (J.3.25.4 - 8). This might occur voluntarily, as by withdrawal of consent, or involuntarily, as by death. Status-loss was regarded as equivalent to death (G.3.153). Dissolution for one partner dissolved the partnership for them all. If those who remained wished to continue as partners, they did so as a completely new partnership. Where a partner voluntarily withdrew his consent and so dissolved the partnership, he might by this very fact become liable to his former partners. If he withdrew fraudulently in order to keep for himself a future acquisition which should, under the terms of the partnership, have become common property, he was required to contribute to his former partners what would have been their share had the partnership subsisted. Even where, without fraud, he renounced the partnership, he was still liable should the others have suffered a loss through his renunciation at an inopportune moment in the partnership affairs (Paul, D.17.2.65.5). Clauses in the partnership agreement purporting to prohibit individual renunciation in principle were void, but a renunciation in breach of such a clause, although the partnership was dissolved, would in some circumstances give rise to liability for resulting loss (Paul, D.17.2.65.6).

(d) Mandate
(J.3.26)

Essentially mandate was the undertaking, by request, of a commission for another person which was both lawful and possible. The contract had its origin in the performance of acts of friendship, which explains why in principle it always remained gratuitous. In developed law a distinction, turning on the question of 'interest', was made between actionable mandate and a non-actionable request or advice. Historically the first enforceable mandates were those involving an interest of the mandator. However, the range of commissions held to constitute mandate was increased. Thus Gaius states that there is a mandate when a commission is given in the interest of the mandator or of a third person. If the commission was purely in the interest of the person receiving the request there was no mandate, as, for example, where A tells B to lend out his money at interest. Difficulty was caused by the case where A told B, not simply to lend out his money, but to make a loan at interest to a named third party, say Titius (G.3.156). Was this distinguishable from the case where someone was commissioned to do something purely for his own benefit? The republican jurist Servius had held not, but the opinion of Sabinus was followed that there was a valid mandate because the commission was also in the interest of Titius; this may be inferred from Gaius' formulation of the reason: 'but for the mandate, you would not have chosen to lend to Titius'. Thus, as J.3.26 pr makes explicit, it was accepted that a mandate might be in the interest both of the person given the commission, the mandatary, and of a third party, or, indeed, of the mandatary and the mandator; only where it was exclusively in the interest of the mandatary was there no contract. The practical result of holding that there was a mandate where A asked B to lend to Titius was that B could sue A if he suffered loss thereby; in effect, A acts as guarantor for the loan to Titius without having to conclude any of the usual stipulatory forms.

As with the contract of partnership, the consensual basis of the contract was given greater effect than in the case of sale or hire. Thus, subject to some qualifications, unilateral withdrawal of consent, whether voluntary or involuntary, as by death, ended the contract. A mandator might revoke the mandate at any time prior to the commission being carried out by the mandatary. Likewise the latter might renounce the mandate and so end the contract, though if he did so in circumstances causing loss to the mandator he was liable for the loss.[53] Death of either party ended the contract with the sensible proviso that should the mandator die, but the mandatary, in ignorance of the death, have carried out the mandate, the former's heir was liable under the contract.

Should the mandatary agree to carry out the mandate upon certain terms, for example to buy a house at not more than a specified price, but fail to observe the terms, as by exceeding the price (buying at less was held to be within the sphere of the original consent), the transaction was deemed to fall outside the scope of the mandate, with the result that the mandatary could recover nothing under that contract. In fact this was the view only of the Sabinians, though Gaius in his *Institutes* gives the impression that it prevailed at his time (G.3.161). However, in a different work, *Golden Words*, Gaius in fact cites and approves the opposing view of

Proculus, also followed by J.3.26.8, that the mandate could be enforced up to the amount specified by the mandator (D.17.1.4).

The essential duty of the mandatary was to carry out the mandate. Where he failed to carry out this duty or carried it out in an improper fashion, he was liable for the amount of the mandator's loss. In classical law, for the mandatary to be liable, it is normally thought that there must have been fraud, some deliberate neglect of duty on his part, a view justified on the ground that mandate was gratuitous. However the matter is not certain. There may have been some circumstances under which the mandatary was liable for non-deliberate fault. Certainly such fault, resulting in extra loss to himself, barred a claim for reimbursement against the mandator. Subject to this point the mandator was required to reimburse the mandatary for expenses directly and properly incurred in carrying out the mandate. His liability did not extend to losses which did not arise directly from the mandate, even if they would not have been incurred but for the mandate, as where the mandatary was robbed while travelling on behalf of the mandator. In Justinian's law the general standard of liability was that of deliberate and non-deliberate fault (see Ulpian, D.50.17.23, often held to be interpolated). These points are not examined in either of the institutional treatments which, apart from observing that the relations of the parties were governed by good faith, typically do not discuss their rights and duties.

Both Gaius and Justinian conclude their account of mandate by emphasising the fact that it is gratuitous. If a payment is agreed the contract is hire or perhaps innominate. In fact under the Empire many persons who performed services under mandate, for example lawyers, doctors, and teachers, did so in return for an agreed salary or honorarium. The position of principle was maintained, however, since such a salary could only be recovered under the extraordinary procedure and not in the action on mandate itself (Ulpian, D.50.13.1; C.4.35.1 (Severus and Antoninus)).

Mandate originated as a contract by which one person commissioned another to perform some act for him, such as representing him in court; one may speak of 'isolated acts of friendship'. In the classical period it also came to be used for the employment of persons performing more continuous services such as teachers or, more significantly, the general agent. The general agent was known in the earlier law but at that time the rights and duties of the parties were regulated by 'uninvited intervention'. During the classical period this was replaced by mandate as the legal basis of the relationship. Certainly, once this had occurred, there would be a case in which the mandatary (the general agent) was liable both for deliberate and non-deliberate loss caused to the mandator.

We have seen that mandate could be used to facilitate arrangements of guarantee (section III(f) above). It was also the means by which an assignment of action was effected. If A had a right of action against B and wished to assign it to C, he gave C a mandate to sue B and released C from the obligation under the mandate to give him what he acquired by suing B. By the time of Justinian this kind of mandate was subject to the special rule that neither death of the mandator nor his express revocation destroyed the mandatary's right to sue B, which previously it did. Also in the law of Justinian, notice of the assignment to the debtor meant that the latter might no longer validly discharge the debt by payment to the original creditor which again, previously, he could.

In one respect mandate was not developed to the extent it might have been. It provided a plank from which a law of agency might have been developed, but progress along this line was only rudimentary. Where A, for example, gave B a mandate to buy something from C, in principle C had no legal relationship with A at all. C was limited to his rights under the contract of sale against B, and equally A was limited to his rights under the contract of mandate against B. To some extent the law, even before Justinian, introduced modifications permitting direct legal action between A and C.[54] By the praetorian institorian action a third person dealing with a free-born *institor* appointed under mandate might sue the principal (mandator) directly, but no direct action was given to the principal against the third person. An important extension of the institorian action was made by Papinian at least in the case where a 'general agent' (*procurator*) was commissioned to borrow money.[55] The creditor was given 'a policy action as if institorian' against the principal (Papinian, D.14.3.19 pr). In neither of these cases was the third party deprived of his right to sue the mandatary.

With respect to an action by the mandator directly against the third party, unless the mandatary assigned his action to the mandator, it again was only in very limited situations, as where he had no other way of protecting his interests, that the mandator could himself bring the mandatary's remedy as a policy action against the third party.[56]

VI. Innominate Contracts and Pacts

To round off the account of contract something should be said of innominate contracts and pacts, which receive only incidental mention in the *Institutes*. Innominate contracts were agreements which did not fall under one of the four accepted categories of contracts, but were thought worthy of enforceability by the praetor. He did not enforce the agreement as such, but provided an action on the case where one party to an agreement had performed his part of the bargain but the other had not.[57] These contracts thus possess the peculiarity that, although constituted by agreement, they become enforceable only on part performance. For this reason they are sometimes (misleadingly) termed the innominate real contracts. The action on the case developed into the good-faith action with a special preface which, notwithstanding their absence from the institutional scheme, heralded the recognition of these agreements as contracts by the state law (see

Chapter 6, section V below). The praetor did not enforce all agreements not falling within one of the classes of nominate contracts, merely those exhibiting an element of *quid pro quo* or what English law terms 'consideration'. To be enforceable the agreement must have been one in which both parties agreed to pay, give, or do something. The possible content of such agreements was infinite, but two important examples to which special names were given were exchange (*permutatio*) and *aestimatum*, an agreement where A supplied B with goods upon terms that B would pay an agreed price by a certain time or return the goods.

Pacts were simple agreements that did not fall within the scope of the nominate or the innominate contracts. Unless incorporated as terms in existing contracts and therefore enforceable by that particular contractual action, in principle, even in the time of Justinian, they were not directly enforceable but could be pleaded only as a defence as, for example, where a pact not to sue had been concluded. Nevertheless, in certain cases the praetor or the emperor made specific pacts directly enforceable. Important examples were *constitutum*, an agreement to pay a debt whether owed by oneself or by another, and *receptum*, an agreement by a shipmaster, innkeeper, or stablekeeper to keep safe goods entrusted to his care. The latter, by Justinian's time, was incorporated into the contract of hire as an implied term imposing liability on the custodian for loss of the goods except in circumstances of an act of God.[58]

VII. Obligations as though from Contract
(J.3.27)

Justinian's account of obligations as though from contract is taken from, but supplements, Gaius' account in *Golden Words* (D.44.7.5). In his *Institutes* Gaius says nothing of the topic. The basis of the classification is negative: it comprises a number of heads of liability which are derived neither from agreement nor from delict. Although not stated, the implication is that the cases included in the title are thought to be more akin to contractual than delictual liability. Notwithstanding the lack of agreement between the parties, each case was in fact probably thought to have certain affinities with one of the recognised contracts.[59]

The most important instance of obligations as though from contract was that arising from uninvited intervention (*negotiorum gestio*; see D.3.5) where one individual, without being asked, though with the intention of being reimbursed (which distinguished this case from gift), intervened in the affairs of another for the latter's benefit. 'Affairs' meant some matter affecting the other's business or property, such as the repair of the house belonging to an absent neighbour which had been damaged in a storm. By means of the counter-action, the person intervening was entitled to recover any expenditure incurred in the interest of the person on whose account he had intervened (the principal), provided it was useful and reasonably incurred. If the intervention was useful when done it was immaterial that it turned out not to have been beneficial; for example, if a sick slave who had been cared for subsequently died (Ulpian, D.3.5.9(10).1). The intervener had to show the highest standard of care; it was not sufficient that he kept to the standard he habitually showed in dealing with his own affairs. This is explicable on the ground that he had intervened, unasked, in the affairs of another.

The principal had the direct action against the intervener with respect to any loss incurred through the latter's fault and to recover anything which he had acquired as a result of the intervention; for example, if he had received payment of a debt owing to the principal. Both the direct action and counter-action on uninvited intervention were good-faith remedies which grew out of praetorian actions on the case.

Where someone requested another to carry out some transaction on his behalf the appropriate liability was under mandate, but in the early classical period the demarcation between mandate and uninvited intervention did not always turn on the presence or absence of a request. The general agent (*procurator*) who had been requested to act on behalf of his principal was liable under uninvited intervention and not mandate, because, although there was an initial general authorisation by the principal, with respect to individual transactions the *procurator* acted according to his own discretion. Another important relationship governed by the rules of uninvited intervention was that between a supervisor and his ward. By contrast the obligations between a guardian and his ward also arose as though from contract but gave rise to the special action on guardianship.

Of the other examples mentioned by Justinian of liability arising as though from contract, the only one that requires specific comment is payment in error (J.3.27.6). Gaius followed a traditional classification which grouped payment in error with *mutuum* as examples of obligation to repay enforceable by the action of debt (G.3.91). However, Gaius himself observes that payment in error cannot properly be classified with *mutuum* once the latter had become recognised as a contract, since the person who pays wishes to discharge a debt, not to contract one. In *Golden Words* Gaius again discusses payment in error, but this time in the section devoted to obligations arising as though from contract (D.44.7.5.3). Justinian follows Gaius' example by retaining a reference to payment in error in the section on obligations contracted by conduct (J.3.14) and reserving a fuller discussion until this title. Nevertheless, one should note that the analogy with *mutuum* is still stressed.

Liability to repay was imposed upon a person who in good faith received payment of a debt in error, but the onus of proof was on the person seeking repayment. He must prove that although the alleged debt had not in fact existed, he wrongly believed that it had; he could not reclaim if he knew there was no debt (Ulpian, D.12.6.1). But what was the position if he paid when in doubt as to the existence of a debt? In classical law reclaim was not possible, even if there was no debt, unless payment was made on the condition that it could be reclaimed were it found not to have been owed (Ulpian, D.12.6.2 pr). In Justinian's law payment in such cases of doubt was treated as a payment in error and therefore could always be reclaimed where there was in fact no debt (C.4.5.11 of AD 530).

J.3.27.7 states that where the defendant was bound to pay double damages if he denied liability which was subsequently established, as under the Aquilian Act, it was not possible to recover a mistaken payment which was not due. This was to prevent a person paying the single amount of money when asked, then reclaiming it on the grounds that it had been paid in error. By raising the action on debt for repayment, the defendant was in effect denying the liability on the basis of which he had paid. The issue of his liability, which he denied, would thereby be considered in this action and not in that on the Aquilian Act, where denial of liability entailed the payment of double damages.

VIII. Obligations Acquired through Others
(J.3.28)

Justinian, like Gaius (G.3.163 - 167a), has a section on the acquisition of obligations through third persons. 'Obligation' in this context means 'right' (the incurring of liability through the contract of a third person is a different matter). Three classes of case are distinguished: (1) acquisition by a single individual through a person within authority; (2) acquisition by two or more individuals through a person within authority; and (3) acquisition through a person not within authority.

With respect to the first class, the rule in Gaius' time was that any right acquired by a slave or son within authority was acquired by the master or father. The one exception, applicable only to sons, was the military fund. By the time of Justinian, sons were placed in an even more favourable position. Not only was the fund 'as though military' treated more or less in the same way as the military fund, but legislation of the emperor Constantine (AD 312-37) vested in the son ownership of property derived from the mother, reserving only a usufruct for the father (C.6.60.1 of AD 316). Justinian extended this rule to all property obtained by the son other than from the father himself, whether the acquisition was through contract or some other means such as legacy or gift.

The second class refers to a situation in which two or more persons own a slave in common. The general principle was that whatever the slave acquired, even through a transaction pertaining only to one of the co-owners, was acquired for them all in proportion to their respective shares in the slave. However should the slave expressly contract in the name of one of them (Gaius and Justinian mention stipulation), only the named person acquired. Gaius records a school controversy on the effect of an acquisition made on the order of one of the co-owners, the Sabinians holding that the order had the same effect as the express naming of that owner by the slave, but the Proculians treating the order as of no effect. Justinian affirmed the Sabinian view (cf. C.4.27.2.(3) of AD 530).

The third class (J.3.28.1) refers to a situation in which someone has been granted a usufruct in a slave, or a free person is possessed in good faith as a slave. The general rule was that the usufructuary or possessor in good faith was entitled only to what the free person or slave acquired from their own labour – where they themselves hired out their services or from the use of capital owned by the usufructuary or possessor in good faith.

IX. Discharge of Obligations
(J.3.29)

Gaius' account of discharge of obligations (G.3.168 - 180) is more elaborate than that of Justinian. The reason is that a number of methods of discharge existed in classical law which had become obsolete by the time of Justinian. In both classical and Justinianic law discharge by payment or performance was the most important mode. Of the other possibilities mentioned by Gaius, two no longer appear in the law of Justinian. These are discharge by an imaginary or fictitious payment accomplished through the performance of a formal act by bronze and scales, and discharge by joinder of issue in a legal action.

As for the other classical modes of discharge, one notes that Justinian resolved a number of doubts expressed by Gaius and that he introduced an important change with respect to novation.

Gaius raised the issue whether a debt might be discharged by the performance or payment of something other than what was owed (substituted performance). He noted that while members of his own school (the Sabinians) held this to be possible provided the creditor consented, the Proculians argued that such alternative performance did not discharge the obligation at law but merely provided a defence of fraud should the creditor sue upon the obligation. J.3.29 pr adopts the Sabinian view.

Verbal release was the most important formal mode of discharge in that, through the mechanism of the Aquilian stipulation which is discussed in detail by Justinian (J.3.29.2) but merely mentioned in passing by Gaius, obligations arising from any kind of contract could be novated in the form of a stipulation and then discharged by the appropriate formula. Gaius' doubt whether an obligation could be discharged only in part through verbal release is resolved affirmatively by J.3.29.1.

Novation was the discharge of one obligation by the substitution of another for it. In early law the contract created by writing could be used to achieve novation, but both Gaius and Justinian discuss it purely in terms of stipulation. It is from the fact that stipulation constituted the principal mechanism of novation that some of its fundamental rules probably derive. The first of these rules is that the content of the debt in the new promise must be the same as the old debt. Stipulation was a contract of strict law. A stipulation intended to extinguish a debt for X amount, however concluded, must itself promise payment of X. If the parties intended X to be extinguished but promised Y instead, in principle an obligation would exist for both X and Y.[60] Yet, on the other hand, there is a rule that 'of two stipulations for the same [thing] between the same [people], one is void'.[61] The application of this rule means that in order to effect a novation, while preserving the content of the original debt, the new promise must also add something new. The clearest example is constituted by a change of parties, as where what is owed by you to Seius is promised by Titius (J.3.29.1). Where the second stipulation is taken from the same debtor, Justinian observes that the new feature might be the addition or subtraction of a condition, a postponement, or a guarantor. As regards the case of the guarantor, Justinian was settling a Sabinian-Proculian dispute (G.3.178). In the case of the addition of a condition in the novating stipulation, Justinian adds that 'the novation happens if and when the condition is fulfilled. The old obligation survives if it fails' (J.3.29.3). It is helpful to look at the background to this statement.

The early history of novation had seen some dispute. The republican jurist Servius Sulpicius had gone further than other jurists in holding that a conditional novating stipulation took effect and extinguished the prior obligation irrespective of whether the condition was satisfied. Gaius rejects Servius' decision (G.3.179). Like Justinian, he states that the novation happens only if and when the condition is fulfilled. The reason is that it is only then that an obligation comes into existence which substitutes for the old obligation. If the condition fails the old obligation survives. However, as an important qualification, Gaius adds that in practice a conditional novation will destroy the old obligation since, even if the condition fails, a defence of fraud or agreement will be available if an action is brought on the original claim which has survived. The reason here is that the parties to the novating stipulation intended that the old obligation should be exacted only if the condition was fulfilled. As an illustration, let us assume that the original obligation is X. A novating stipulation is taken in the form, 'I promise X if it rains before Tuesday'. It does not rain, and therefore in law the original obligation X survives. However, it is clear from the terms of the stipulation that the parties intended that X should be exacted only if it rained, which it did not, and it is for this reason that Gaius allows the said defences against a claim for X. Justinian does not mention the defences. In reproducing only the first part of Gaius' statement, namely that the prior obligation subsists if the condition of the novating stipulation fails, he implies that, for his law, no defence of fraud or agreement was available.

Servius Sulpicius also held that a stipulation taken from a slave was a valid novation of an existing debt and hence ensured loss of the claim, since the slave himself could not be sued. Gaius rejects this view and is followed by Justinian, who says that it is as if there had been no subsequent stipulation at all (J.3.29.3).

Gaius says nothing of the requirement that there be an 'intention to novate' discussed by J.3.29.3a. Clearly there had sometimes been doubt as to whether a stipulation was intended to replace and so novate an existing obligation, or whether it was to be construed as the addition of an obligation to that already existing. Hence Justinian provides that a stipulation should not extinguish a prior obligation unless it expressly states that the intention of the parties is to novate (C.8.41(42).8 of AD 530).

Among modes of discharge Justinian also includes withdrawal of consent by both parties to an obligation concluded by mere agreement. In the case of partnership or mandate, unilateral withdrawal sufficed, provided in the latter case that nothing had been done in fulfilment of the contract. Although not mentioned by Gaius, this mode of discharge was known in classical law.

X. Obligations from Delict
(J.4.1 - 4)

(a) Introduction

Delictual obligations arise from wrongdoing and are sanctioned, for the most part, by penal remedies. The function of such actions was to punish the wrongdoer, and should be distinguished from that of restorative actions (J.4.6.16; see also Chapter 6, section VI below), including those arising from contract, which was to compensate the plaintiff for his financial loss. From this difference in function there resulted a difference in structure between penal and restorative remedies which is most clearly seen in the characteristics that each exhibited. However, as the law developed it is thought that the function of many penal actions also became that primarily of indemnifying the plaintiff for his financial loss. In such cases, to varying degrees, these actions lost the penal characteristics which they had had in their early history. Many scholars nevertheless still distinguish them from purely restorative remedies on the ground of their penal 'nature', which denotes their origin as a pure penal remedy, notwithstanding the fact that they now fulfil a compensatory 'function', as evidenced by their new characteristics which are those usually found in restorative actions.[62]

The clearest penal characteristic which a remedy might have is multiple damages, as in the theft actions. However, multiple damages is not a feature of all penal actions, and where it is not a feature a penal action may be identified on the basis of other characteristics which it possesses. Traditionally these characteristics are explained by assuming that in origin penal actions were a substitute for private vengeance. Where more than one person had committed a wrong, vengeance was meted out to all. Consequently, a penal action lay in full against all joint wrongdoers and if one paid his penalty the others were not released from paying theirs. The purpose of the action was punishment and hence the plaintiff was able to recover in full from each offender. Vengeance is personal, which means that it was permissible only against the wrongdoer. This is thought to explain why penal actions did not lie against the wrongdoer's heir. Penal actions were also noxal and generally only available for a period of one year after the commission of the wrong. In all these respects restorative actions were usually quite different; for example, if one party had paid the damages, any others who were jointly liable were released. Similarly, because the function of such remedies was to compensate the plaintiff for his loss, not to punish a particular individual, it was immaterial that the defendant had since died and the action could therefore be brought in full against his heir.[63]

(b) Theft
(J.4.1)

Justinian's definition of theft (J.4.1.1) corresponds, with one important variation, to a definition in the *Digest* attributed to Paul (D.47.2.1.3). It neatly pinpoints both the main requirements for the commission of the delict and the range of circumstances which it covers. As will be seen, these two matters are probably not unconnected. The phrase 'handling ... with fraudulent intention' expresses the physical and the mental conditions normally necessary for theft (*contrectatio* and *animus furandi*), whereas the clause 'in relation to the thing itself or even its use or its possession' refers to the fact that theft may consist, not merely in the removal of another person's property, but in the misuse of another's object which one had properly received, and sometimes even in the removal of one's own property. Thus, for example, it was theft where a person who had received physical control of another's object under a contract of deposit made use of it, and, in *commodatum*, where the borrower put the object to a use not authorised under the contract (J.4.1.6).

Theft of one's own property occurred where the owner removed his object from someone who presently had a real right over it, such as a pledge creditor (J.4.1.10). The real right was usually, but not exclusively, possession, which explains why this case is often referred to as theft of possession. Similarly, unauthorised use of property, as described above, is often called theft of the use. Added to these is theft of the object itself, giving a threefold classification of the delict. The origin of this classification, which was not Roman, has been traced to a tendency to misinterpret Justinian's definition of theft by reading the clause, 'in relation to the thing itself or even its use or its possession' as dependent on the word 'handling' (*contrectatio*).[64] This leads to the idea that theft is constituted in three separate ways, by the 'handling' either of the object itself, its use, or its possession. Yet this is absurd, because how can one have a handling of the use or possession of an object? In fact, Justinian's definition lacks the words 'with a view to gain' which are contained in the definition found in the *Digest*, and it is on this phrase that the clause, 'in relation to the thing itself ... or its possession' should depend. There is therefore only one form of theft, namely that of the object itself. The intention of the thief, however, may be to make a gain either from the object, its use, or its possession.

'Handling' is unlikely to have been a sufficient requirement of theft in the earliest law. Generally it is thought that at the time of the *Twelve Tables* the physical element of theft was actual removal or asportation. This conclusion is reached mainly by inference from the etymological connection between the word *furtum* (theft) and the verb *ferre* ('to carry off'), a point which is reflected in J.4.1.2. This raises the question whether the idea of carrying off merely says something about the central image of the wrong, which is that of a carrying off of property, or whether it expresses what was in fact an essential requirement of theft.[65]

If asportation was a necessary requirement in early law, certainly by the end of the Republic it was beginning to be replaced by 'handling'. This was an important change because its effect was to make a person liable to the heavy penalties of theft if merely caught touching another person's property in circumstances indicating that he was preparing to appropriate it. However, the reason for the change was not to bring attempts within the category of theft, but rather is suggested by the context in which 'handling' is introduced in the institutional accounts. Gaius (G.3.195), whose treatment is closely followed by J.4.1.6 - 8, remarks that theft is committed, not only by removal of another's object, but also by the 'handling' of it, which he then illustrates with the cases of deposit and *commodatum* already mentioned. 'Handling' was clearly found to be convenient as the general term with which to express the physical requirement of theft. This was because it covers not only cases of actual removal, but also cases where there had been no removal in the sense of an unauthorised taking away but where some other kind of misuse or misappropriation of another's object had occurred, as in the example given of deposit, where the thief in fact holds the property at the request of the owner.[66]

One can see from Paul's statement, adopted by Justinian, that 'handling', certainly by the late classical period, had assumed a central role in the definition of theft; so much so that some jurists seemingly treat it as an essential element of the delict (Ulpian, D.47.2.52.19). This creates a difficulty because there are a number of well-known cases in which there appears to be no 'handling' by the person who is made liable for theft. For example, a person was held liable who maliciously summoned a mule driver to court with the result that the mules were lost (*Veteres*, in Paul, D.47.2.67(66).2). A variety of possible solutions to the problem has been offered.[67] Most of these explanations assume that there was a common attitude amongst the jurists, not only concerning the precise meaning of 'handling', whether, for example, it entailed an actual physical touching or not, but also concerning its position as an element of theft. However, it is more probable that differences existed between the jurists. For instance, one jurist might regard 'handling' as an essential element but construe it widely, while another interpreted it narrowly but was prepared to treat as theft cases in which it was not present. A reason for doing so might simply have been that traditionally the particular case before the jurist had always been treated as theft.

In contrast to the pre-eminent position assumed by 'handling', it is noteworthy just how many terms were used to denote the mental requirement of theft, though this is not reflected in either of the institutional accounts. The mental requirement is expressed differently depending on the context. The general tendency was for the jurists to examine more closely whether a specifically theftuous intent was present the less this was evident from the nature of the conduct. Thus, for example, where property had actually been removed from the owner, this fact alone was often sufficient to make it clear that theft was the motive, in which case the wrongdoer might merely be said to have acted deliberately or without the owner's consent. However, the motive is not clear in the case where lost property is picked up and taken away, because such behaviour does not reveal whether or not the finder intends to keep it. Hence, in this instance what determines whether the taker commits theft is his intention to profit, evidenced by a failure to seek and return the object to its owner.[68] A general expression sometimes found in the texts to designate the mental criterion is *animus furandi*, 'an intention to steal'. As a criterion this is unhelpful since the formulation already presupposes that there is theft. One simply has here a convenient juristic shorthand in which the intention to appropriate (the criterion) has been telescoped with the legal result (theft).

The standpoint taken by Gaius (G.3.195), followed by Justinian (J.4.1.6), is that the mental requirement of theft was furnished by someone handling property without the owner's consent. We should note that for the normal case it was not thought necessary to state that the wrongdoer knew that the owner would not consent; such a mental state was implied from the clear circumstances of the case. However, in the context of *commodatum* the point is made that unauthorised use of the object borrowed counts as theft only if the borrower knows that the owner would not consent. If he believes that the owner would have consented had he known, this is not theft. It is necessary to state the rule in this context because the borrower holds and uses the object with the permission of the owner. It is therefore not necessarily obvious merely from the borrower's further use, albeit that this is unauthorised, whether he has the requisite mental state to be guilty of theft. He may honestly believe that the owner would have consented to the further use. However, he must have reasonable grounds for this belief. From the examples given in J.4.1.6 it seems that he was deemed to know that the owner would not consent, and was therefore guilty of theft, where he put the object to a use which might ordinarily have resulted in its being permanently lost to the owner, such as where he took borrowed silver abroad or a horse into battle. In effect the borrower is guilty of theft only where his unauthorised use amounts to a taking.

Gaius (G.3.198) considers an additional refinement concerning the mental requirement of theft. Even though a person believes he is handling something against the will of the owner, there is still no theft if in fact the latter had consented to the handling. This leads Gaius to pose the following problem: what is the position where A solicits B's slave to steal from B and the slave reports the matter to B, who allows him to take the things to A with the object of catching A in the act of theft? Gaius, following the logic of his proposition, holds that there is no theft since the owner consented to the handling; nor does the action for corruption of a slave lie since the slave in fact has not been corrupted.

J.4.1.8, on grounds of policy rather than logic, reverses this decision and settles the classical controversies by making A liable both for theft and corruption of the slave. The giving of the latter action is remarkable; what, for example, would be the measure of damages?

J.4.1.18(20) also puts a situation illustrating the particular relevance of intention. Could someone under the age of puberty commit theft? The answer was that he might if he was sufficiently close to puberty to be able to understand that what he was doing was wrong. In other words the general test of capacity laid down here is that a person could not be liable for theft unless he was generally able to distinguish between right and wrong. Liability on the part of a small child or a mentally deranged individual was therefore excluded.

J.4.1.3 - 5, which rely heavily on G.3.183 - 194, discuss the various kinds of theft, noting that some are now obsolete. In Justinian's law theft was either manifest or non-manifest, the latter class covering both the initial theft and any subsequent receipt of the goods by a person knowing them to have been stolen. Gaius gives a fuller statement of the early law, explaining that in the case of manifest theft the *Twelve Tables* imposed the death penalty if the thief was a slave, and provided that a free person should be scourged and then handed over by the magistrate to his victim as a sort of slave, his exact legal status having been a matter of dispute amongst the early jurists. Subsequently these draconian penalties were replaced by the praetor with an action for four times the value of what was stolen. The praetor preserved the twofold pecuniary penalty given by the *Twelve Tables* for non-manifest theft.

From Gaius' account it is clear that in classical law there was some disagreement over some of the circumstances under which a theft counted as manifest. He himself is inclined to restrict it to cases in which the thief is caught actually in the act or at least in the place where the theft had been committed. But J.4.1.3 overrules Gaius on this point and, following the opinion of Julian and Ulpian (D.47.2.3.1, 2), extends the notion to the case in which the thief is seen or caught prior to his reaching his destination, subject to the qualification that it is the destination he intended to reach on the day of the theft (Paul, D.47.2.4). The reference to 'seen' is important since it suggests that Justinian modified the classical view which refused to treat 'seeing' the thief as equivalent to 'catching' him (Ulpian, D.47.2.7.1, 2).[69] As Justinian observes, all cases of theft which were not manifest were non-manifest.

The difference in their penalties suggests that manifest was seen as more serious than non-manifest theft and that it was for this reason that they were distinguished. But Jolowicz[70] draws attention to the fact that in early law there was also a procedural difference between the two forms of theft. Only in the non-manifest case was there a trial. The reason must have been that it was unnecessary for manifest theft because the guilt was clear. If proof of theft was furnished only by catching in the act, perhaps the true penalty for theft was that of the manifest thief. The less severe pecuniary penalty for non-manifest theft would have been due to doubts as to the adequacy of the proof in this case. However, after the introduction of the pecuniary penalty for manifest theft the procedural difference ceased to exist because it now also gave rise to a trial.

It is not mentioned by Justinian, but at the time of the *Twelve Tables* there existed a formal search in which, according to Gaius (G.3.192 - 193a), the person searching had to do so naked but for a loincloth and carrying a dish.[71] In the event of the stolen property being found on a man's premises he was treated as a manifest thief. What is mentioned by J.4.1.4 is an informal search with witnesses, which in early law existed side by side with the formal search and which gave rise to the action for theft by receiving, for three times the value of the stolen property found on the premises. Why should two such searches have been available? Daube[72] shows that ordinarily a person searching for a stolen object would use the informal method. Only if this was resisted by the householder would there be recourse to the more formal search. Gaius observes that someone who will not allow you to search with your clothes on is hardly likely to let you do so with them off, but Daube's response is that at the time of the *Twelve Tables* the formal search was of a semi-sacred nature; it had 'all the power and prestige of secular and religious authorities to support it' and 'could hardly be resisted'.[73]

The formal search was obsolete by the time of Gaius, having been replaced by the praetor with an action for theft by prohibition, which now lay where the informal search was resisted. By his mere act of refusal of the informal search the householder was a manifest thief. This is shown by the fact that, like the action for manifest theft, the action for prohibition lay for a fourfold penalty and was a praetorian substitute for the old formal search, which itself had resulted in the householder being dealt with as a manifest thief.

An interesting feature of the action for theft by receiving is that it lay against the person on whose premises stolen property was found even if he was not the thief. Relief was given to him only by means of the action for planting which also lay for a threefold pecuniary penalty. Originally this action possibly lay against anyone from whom he received the object, even if that person was not the thief, but both Gaius and Justinian state that liability arose only when the planter intended that it be found with the receiver rather than him. Note that condemnation in the action for receiving was the only qualification for bringing the action for planting.

All these remedies (including an action for theft by retention, not mentioned by Gaius; see J.4.1.4) appear to have been in use in the classical period but were obsolete by Justinian's time. The private searches conducted by the victim were replaced by searches by officials, possibly the staff of the magistrate to whom the offence had been reported. Knowledge that the goods were stolen was now required, the person on whose premises they were found being a non-manifest thief.

J.4.1.9 - 10 deal with the kinds of object of which theft is possible, not giving an exhaustive list but mentioning only some unusual cases. However, neither Gaius nor Justinian say anything about land. Notwithstanding the opinion of some jurists (Ulpian, D.47.2.25 pr), it was excluded from the sphere of theft. The reason was that it could not be carried off and therefore fell outside the central conception of what constituted this delict, namely that of carrying off other people's property.

J.4.1.11 deals with the liability of an accomplice, the difference between principal and accomplice being that only the former 'handles' the goods and thereby is a thief proper. Since both principal and accomplice were liable in the action for theft to the same penalty, in practice there was no difference between them, though there were circumstances where a person was liable as an accomplice even though the principal was not (J.4.1.12(13) - (14)). Gaius (G.3.202), followed by Justinian, stresses the importance of intention in determining the liability of the accomplice. If one person's act enables another to steal and yet the former has not intended theft and has not acted in collusion with the thieves he may, on the ground of his fault, be liable to an action on the case, but there will be no theft action against him. J.4.1.(12) states that there is no liability as an accomplice unless actual physical assistance is given; hence one is not liable in theft merely for advising or encouraging someone else to take another's property. It is probable that developed classical law was different on this point and held a person liable as an accomplice either where he gave physical assistance or where he gave advice or assistance in the plotting of a theft.[74]

Normally the person entitled to bring the action for theft was the owner, but there were cases, detailed in J.4.1.13(15) - 17(19), in which persons other than the owner were allowed to do so on the grounds that they had what both Gaius and Justinian refer to as an interest in the safety of the thing. In identifying the nature of the interest necessary to ground the theft action, scholars have found it convenient to distinguish a 'positive' or 'negative' interest. Generally a person with a positive interest had the action on theft concurrently with the owner, whereas a person with a negative interest had it to his exclusion. Standard examples of persons with a positive interest, who normally had a real right over the object, were the pledge creditor and the usufructuary. Both had an interest in the object independent of that of the owner and hence were given the theft action with respect to that interest. Thus the pledge creditor had the action to protect the value of his security interest, which was the amount he had lent to the debtor, but the debtor himself retained the action for the difference between the amount of the sum borrowed and the value of the object. In the case of usufruct, a formula specified the separate interests of usufructuary and owner and the action was given to each to the appropriate extent.

Certain contracts of hire, those with the laundryman and tailor, and that of *commodatum*, illustrate the workings of negative interest. The general principle was that the person with control of the object under the contract was exclusively entitled to the theft action because he was liable for the loss of the object under the contract. One limitation was that his insolvency, preventing recovery of damages by the owner, effectively deprived him of his negative interest and left the owner with the action. A more difficult point concerns the extent of liability of the laundryman, tailor, or borrower under their respective contracts. For classical law it is generally assumed that they had an insurance liability, which meant that they were strictly liable for the loss of the goods in all circumstances except those constituting an act of God. In Justinian's law this had been relaxed to liability on account of failure to exercise the care of a reasonable man. On this view there would be an important difference between classical and Justinianic law in the operation of the rules on entitlement to the theft action. In classical law the laundryman, tailor, and borrower had the action even where the object was stolen without carelessness on their part, since they remained liable under the contract. In Justinian's time, they were not entitled to the action in these circumstances because they were no longer contractually liable. However, it is questionable whether such a sharp distinction existed between classical and Justinianic law. In both periods the matter probably depended on the operation of presumptions. Where the object was stolen from, say, a borrower, the principle operated that he was at fault in failing to keep it safe, and he was therefore liable under the contract. But this was not an irrebuttable presumption. Should he be able to show that all due care had been taken he would disprove fault, and, because he was not liable under the contract, he would not be entitled to the theft action.

In classical law the owner had the theft action only where the person holding the object under a contract of hire or loan was insolvent. It is clear from J.4.1.16(18) that as regards loan Justinian saw this as unsatisfactory. He therefore changed the law by giving the owner a choice, either to sue on the contract, in which case the borrower had the theft action, or to bring the action on theft against the thief, which resulted in the borrower being released under the contract. In exercising his option the lender might take into account the borrower's solvency, but his legal position was no longer dependent upon it. Justinian does not specify what he found unsatisfactory in the classical law position. However, one danger was that the borrower might have become insolvent after he had sued the thief but before the owner brought the action on the contract against him. The action against the borrower was worthless in such a case, yet the owner could not sue the thief for a second time. Alternatively, if the owner was first to sue on the contract, it was inconvenient that he should then have to raise the theft action against the thief because the borrower had been found to be insolvent. Justinian probably intervened in the case of loan and not hire because it was a gratuitous contract from which the owner derived no benefit.

The depositee did not have the theft action because, being responsible only for his own wilful default, he was not liable on the contract in the event of losing the object by theft (J.4.1.17(19)). Even if he was liable on the contract he still was not entitled to the action for theft since by general principle a person was not allowed to profit from his own wilful default as he would do if he was given this remedy.

An important procedural point is made in J.4.1.19(21). The purpose of the theft action was purely to punish the thief, which it did by the imposition of a penalty. As regards the quite separate issue of compensation for the loss of his property, the owner had a choice of one of two remedies. The vindication lay to recover the object itself from whoever was found in possession of it, whether he was the thief or not. The action of debt, unlike the vindication, did not make an assertion of ownership but merely stated that the thief (or his heir), whether he had the object or not, was under a duty to convey it to the owner. This was a particularly valuable remedy where the object could not be traced or was known to have been destroyed, or where the plaintiff was no longer its owner, for example, because of the operation of the rules of accession in favour of another, even the thief himself. In effect, by this action the plaintiff sought damages for the non-return of the object; it did not succeed where the object had in fact been restored to him.

(c) Things taken by force
(J.4.2)

Liability for robbery was introduced by the praetor Lucullus in 77 BC to cope with problems of serious political disturbance and violence which followed the Social War. The edict granted a special remedy where someone's goods had been forcibly taken, which became known as the action for things taken by force. The relationship of this remedy with one introduced around the same time to penalise loss (*damnum*) caused by the activities of armed bands is not clear, the main questions being whether the edict on loss was so comprehensive as to have included robbery within its terms, or whether each was originally treated as a separate wrong giving rise to its own action on the case.[75] Gaius mentions robbery in his *Institutes* (G.3.209), which were devoted to the state law, because it was an aggravated form of the state-law wrong of theft.

The action for things taken by force lay for a period of one year for four times the value of the object forcibly taken, and thereafter for its simple value. The 'fourfold' therefore represented a penalty of three times the value plus compensation for the loss of the goods. However, in classical law the accepted view became that the 'fourfold' was a pure penalty, as shown by the fact that to claim compensation the plaintiff was allowed additionally to bring either the action of debt or vindication (G.4.8; Julian, in Ulpian, D.13.1.10.1). Justinian reverted to the action's early conception and described it as being neither purely penal nor restorative but hybrid in character with the result that the above-mentioned restorative remedies were again excluded (see Chapter 6, section VI below).[76]

The action for things taken by force was always a bar to the action for theft. However, before the expiry of the period of one year, more was recoverable with the action for manifest theft plus the vindication and action of debt than by this action. It is likely to have been this factor which in classical law led to the action for things taken by force being treated as a pure penal remedy, because why should such a flagrant wrongdoer be treated more leniently than a thief? Yet at this time, in practice, the plaintiff will seldom have had a choice between these two remedies because the use of force will have precluded the wrongdoers from being caught within the terms of manifest theft. Once Justinian had extended the notion of manifest theft to include 'seeing', there would often have been a choice between the two remedies, but, because he had also decided that the action for things taken by force should be hybrid in character, generally it will have been more advantageous to bring the action for manifest theft. In circumstances where the plaintiff brought the action for non-manifest theft he was able also to bring that for things taken by force for any additional amount recoverable under the latter. The action for non-manifest theft was perpetual, and therefore it will have been the preferred action after a year had passed, because the action for things taken by force in classical law would not then lie at all, while in Justinian's time it would only lie as a restorative remedy for the simple value of the property.

J.4.2.2 says that any person who had an interest in the object not being carried off was entitled to the action. On the basis that the action is given to the depositee, it is usually argued that this interest was construed less strictly than that necessary to ground the action for theft. Such an inference is unjustified since the depositee referred to has assumed an additional liability under the contract for non-intentional fault. In effect he has a negative interest in the goods.[77]

In general the normal conditions of theft applied; thus, there must have been a 'handling' (*contrectatio*) of moveables effected with wicked intent (*dolo malo*), which in this context should be taken in the sense of an intent to deprive another of his property. In practice, a mere 'handling' of the goods is unlikely to have constituted robbery, as it might theft, since the exercise of force will nearly always have resulted in the property having been carried off or at least moved.

The remedy is based upon the fact that force has occurred. 'Force' is to be understood not just in the sense of physical violence, but of 'open or flagrant' as opposed to 'secret or stealthy'. A display of force, the overawing or intimidation of the owner or householder, even the possession of an offensive weapon by a single individual acting on his own, would constitute 'force' for the purpose of the action, even though no one was actually physically attacked or injured.

The common element of force explains the inclusion of another wrong beside robbery in this title. Clearly people resorted to self-help and exercised a display of force in recovering property to which they claimed they had a right. Such an act was neither theft nor robbery since there was no intention to deprive another of his property. Nevertheless, to discourage such acts of self-help where no recourse had been made to legal process, certain imperial pronouncements were passed imposing the penalties described in J.4.2.1. A pronouncement of the emperor Marcus Aurelius (AD 161-80), which applied only to moveables, provided that it was 'force' to take that to which one claimed to be entitled without going to court, and that in such a case the claim was forfeit (Callistratus, D.4.2.13; D.48.7.7). A pronouncement of AD 389 (C.8.4.7) supplemented this enactment by extending its rules to the repossession, without authorisation of a court, of any property, including land, and further provided that should what was taken in fact belong to another, it and an equivalent in value had to be paid to him.

(d) The Aquilian Act
(J.4.3)

The Aquilian Act was a plebiscite, traditionally thought to have been enacted around 287 BC. It contained three sections, the first and third of which dealt with the delict of 'wrongful loss' caused by damage to property.

The origin of this delict was therefore statutory and the remedy which was introduced, the action on the very words of the Aquilian Act, was one of state law. The second section also provided a remedy for loss, but in this instance not loss caused by damage to property, but by an additional stipulator who fraudulently discharged a debt due to the stipulator. Gaius (G.3.215 - 216) notes the terms of the second section but implies that it was already unnecessary in his day because the stipulator would have the action on mandate in such circumstances. J.4.3.12 observes that the second section had become obsolete.

The odd position of section two between the two provisions concerning damage to property has been explained by the assumption that preceding the Aquilian Act there existed some statute which had issued rules on both the killing of a slave or animal and the case of the fraudulent additional stipulator. When the Aquilian Act was introduced it reformed the section on killing by removing what had been the fixed pecuniary penalties available in that case, and introduced a further provision on damage to property in its third section. As is common in early legislation when further provisions are introduced into an existing code, the new section was simply added on at the end.[78]

In D.9.2.1 pr Ulpian says that the Aquilian Act derogated from every preceding act which dealt with wrongful loss, whether it be the *Twelve Tables* or any other statute. It is not entirely clear what he meant by the word 'derogate', or indeed whether he had in mind other statutes in addition to the *Twelve Tables*. The latter certainly contained miscellaneous laws on aspects of what might be called damage to property: for example, the provision which established a penalty of 150 asses for breaking the bone of a slave. However, this is treated as an instance of the delict of contempt in the *Twelve Tables* (J.4.4.7) but as wrongful loss by the time of the Aquilian Act. Therefore, unless, as has been forcibly argued by Kelly,[79] the Aquilian Act itself marked the first stage in the process by which these two wrongs became distinguished, the likelihood is that preceding it there must have existed legislation in which the separation between the delicts was effected.

Our detailed sources on the Aquilian Act date from over 400 years after its introduction. It is therefore perhaps not surprising that some scholars have argued that its provisions at its introduction were not in fact as wide-ranging as the classical law presents them. This doubt has centred on section three in particular.

Gaius (G.3.210) gives what is generally, though not universally,[80] thought to have been the original scope of section one. It provided an action on the very words of the Aquilian Act against a person who wrongfully killed another's slave or livestock-quadruped for the highest value which that object had held in the previous year. J.4.3.1 adds a brief account of the kinds of animal falling within the scope of the section, noting that it always excluded wild animals and dogs but that it was subsequently interpreted to include pigs. For him livestock were those animals which could properly be said to graze, which perhaps explains why he does not mention camels and elephants even though they also came to be included under the section (Gaius, D.9.2.2.2).

Justinian's account of section three (J.4.3.13) was copied from Gaius and was the account that was current in the classical period. It provided an action on the very words of the Aquilian Act where a person wrongfully wounded a slave or any kind of animal, or killed an animal not covered by section one, or damaged or destroyed an inanimate object. Just how comprehensive this provision was thought to be is shown by Justinian's statement that it provided for all loss not covered by section one. However, some modern scholars believe that section three, as originally enacted, could not have had so extensive a scope. The fundamental reason for supposing an original restriction is that it appears to allow as damages the highest value held by the object within the 30 days preceding the wrong (J.4.3.14, 15). It is argued that this would produce an absurd result in the case of a slight wound or trifling damage. In order to deal with this difficulty a number of reconstructions of the original section have been suggested. Two of the best known are as follows: it was limited to the complete destruction of inanimate objects alone, in view of which the recovery of the full value is reasonable;[81] or it was limited to the infliction of a wound on a slave or animal, damages being not the highest value of the object but the loss to the owner determined by reference to the 30 days after the wound, by which time its full consequences will have become apparent.[82]

The view taken here is that all such interpretations are too restrictive. No sufficiently strong argument has been advanced to show that the scope of this section was not as wide when introduced as the jurists said it was. As regards damages, the answer is to be found in the words contained in section three describing the acts causing loss for which a remedy was first given: to burn, to damage, and to break. They express damage of a very serious kind to an object. If the section originally contemplated a remedy only for serious physical damage it was not unreasonable to stipulate as compensation the highest value of the object within the previous 30 days. Only when the section was extended to cover cases of less serious injury did the damages clause become inequitable, which in turn led to its being altered by the jurists.[83]

J.4.3.2 - 8 examine the meaning of the word 'wrongful' (*iniuria*) which appears in both sections of the act and in the name of the delict itself. At this level it simply expresses the fact that for there to be an Aquilian action the defendant must have acted wrongfully. Yet we are not told what amounted to wrongful conduct. Ulpian in D.9.2.5.1 defines 'wrongful' as what is done 'without justification', as where one is at fault in killing. Traditionally it has been thought that 'without justification' and 'fault' were separate criteria by which liability was established and that the mention of both is to be explained historically.[84] When the act was first introduced it penalised persons who had acted 'without justification'. What this meant was that a defendant was always liable unless he could show (the onus of proof being on him) that he was exercising a right when he caused the loss in question. An obvious example is killing in self-defence. On this view something very like strict liability was imposed by the act since the defendant could escape only by proving that his actions fell within one of a limited class of legally recognised rights. On the other hand, in circumstances where he could show that he was exercising such a right he was absolutely free from liability. By the Principate the jurists had replaced this regime of strict liability by one based on deliberate or non-deliberate fault (*culpa*). Fault served either to restrict or to broaden the earlier scope of liability by enabling a defendant who had acted without a right to show that he was not to blame for the loss (Paul, D.9.2.28), or, alternatively, where he was to blame, by treating what was previously lawful as actionable (Paul, D.9.2.31).

Culpa means fault in the sense of personal blameworthiness, which can include fraud. When it is used in contrast to fraud it expresses various kinds of non-deliberate fault. Examples given in the texts are failing to take necessary precautions, use of excessive force, and failure properly to exercise a skill. Although many of the factual situations held to constitute *culpa* are examples of carelessness (which itself should be distinguished from the modern legal concept of negligence), *culpa* does not mean carelessness. There are cases of non-deliberate fault, such as brutality, which cannot be classified simply as acts of carelessness.[85]

Gaius (G.3.211) suggests that 'wrongful' in his day was understood exclusively in terms of fault. Justinian's treatment is therefore noteworthy because in J.4.3.2 he reverts to its early meaning of 'without justification'. Thus he says that there is a right where someone kills a robber (slave) in self-defence, though one should note the qualification expressed in the phrase 'if he could not otherwise escape danger'. In J.4.3.3 and subsequent paragraphs one expects to find examples of the later interpretation in terms of fault. Some of the examples (J.4.3.6 - 8) which are drawn from classical texts are of this kind. However, those given in J.4.3.4 and 5, although they are framed in the language of fault, look more like situations which turn on the presence or absence of a right. Thus the liability of the javelin thrower is made to depend upon whether he has a right to be practising in the place in question, and that of the pruner at least in part upon whether the passer-by has a right to be where the pruning is taking place. Use of the language of fault suggests that Justinian is reintroducing as a test of liability the presence or absence of a right in the guise of a presumption of fault. Where the defendant has a right to perform the act which results in loss, there is a presumption that he has not been at fault which is apparently irrebuttable; equally where the slave killed has been in a place where he had no right to be, there is a presumption, again irrebuttable, that there is no fault on the defendant's part.

Justinian, as indeed the classical jurists, discuss the question of fault in considerably more detail than that of causation. The reason is that it was easier to decide difficult cases by pinpointing the incidence of fault than by asking who had actually caused the loss. A well-known decision from the republican jurist Mela (D.9.2.11 pr) may be instanced. A barber was shaving a slave near a place in which some people were playing a ball game. A person hit the ball too hard with the result that it struck the barber and caused the razor in his hand to cut the slave's throat. Instead of asking, 'who caused the death of the slave?', the jurist decides that liability rests with whoever has been at fault.

To be liable, not only must the defendant have acted wrongfully, but he must also have 'caused' the loss for which he was being sued. Problems of causation are treated briefly in J.4.3.16 which summarises the results of an historical development. An action on the very words of the Aquilian Act lay on account of killing only where the defendant had killed directly, that is, by his bodily force. At first this meant where there had been actual physical contact between the parties, as where the defendant killed with his hand or with a weapon held in the hand. But, in time, the jurists construed what amounted to killing less strictly; thus they came to view it as including the case where A killed B by dropping something on him. The furthest they went in this direction was to hold that where A pushed B into a river, so causing B's death through drowning, he still killed directly and was therefore liable in the state-law action on the very words of the Aquilian Act (Celsus, in Ulpian, D.9.2.7.7). However, where A pushed B who fell against and killed C, A was not held to have killed (D.9.2.7.3). This was too indirect.

A major development took place when the praetor came to provide a policy action in cases where death was caused indirectly. These actions were different from the state-law remedy but nevertheless modelled on it, in the sense that they incorporated certain of its special features: for example, damages being assessed at the highest value within the past year and their doubling in the event of conviction following upon denial of liability. But even after this particular development had been made, not all cases of indirect causation were brought within the scope of the remedy. In the words of the jurist Celsus, for a policy action to lie the defendant must have 'furnished a cause of death'. Examples are given by Gaius (G.3.219) though he does not use the language of furnishing a cause of death: shutting up and starving to death another's slave or animal; driving another's animal so hard that it drops dead; persuading another's slave to go down a well or climb a tree with the result that he falls and is killed.

There was not always unanimity between jurists as to whether a particular set of facts should give rise to the action on the very words of the Aquilian Act or to a policy action. We have seen that causing a slave to drown by pushing him into a river came to be regarded as a killing, yet Gaius, while admitting that this might be so, includes it in a list of cases remedied by the policy action. Justinian favours the view that it is a killing.

Section three also underwent both juristic and praetorian development. When introduced it was restricted to cases of serious damage to property, expressed by the verbs 'burn' (*urere*), 'damage' (*rumpere*), and 'break' (*frangere*), which describe the nature of the act for which an action on the very words of the Aquilian Act was originally given under the section. But, in time, the jurists became concerned to bring within its scope cases where there was loss but the physical damage was slight. They did this on the basis of the word *rumpere*. The first stage was to construe it less strictly so as to include cases where the injury was minor, as where someone cut the skin of a slave or gave him a bruise. However, the most important stage, associated in particular with the jurist Celsus, occurred when they read *rumpere* as *corrumpere*, 'to spoil'. The result was that the section became applicable to cases where there was no actual damage but merely a physical alteration, as where wine was turned into vinegar. As is shown in J.4.3.13, 'spoil' is such a wide concept that it absorbed the other three words of the section and the Aquilian action henceforth became applicable to all cases where one could speak of any sort of damage to, or alteration of, property.

Under section three the action on the very words of the Aquilian Act was also available only where the defendant caused loss directly. As under section one, this remedy was supplemented in time by a policy action where the loss was caused indirectly. There is, however, a further difficulty which is raised in the context of section three. A second requirement for the availability of the action on the very words of the Aquilian Act, besides the requirement of directness, was that the loss must have been caused by actual damage to property. If Justinian is to be believed, this condition also applied to the policy action. Thus, both the above remedies were unavailable in the case where someone out of pity freed another's slave from chains and let him escape because, not only had the defendant acted indirectly, but there was no evidence that the slave had been physically harmed, even slightly, so as to come within the understanding of *corrumpere*. In this case where neither of the original two conditions of the act were met, Justinian gave an action on the case on the grounds that the owner of the slave had nevertheless suffered loss.

We see that J.4.3.16 seemingly makes a threefold classification of actions arising under the act depending on the facts. This differs from Gaius (G.3.219), who never mentions an action on the case; and indeed both of these institutional works differ in turn from the juristic commentaries contained in the *Digest*, which use the terms 'policy action' and 'action on the case' interchangeably. Justinian's threefold classification cannot therefore strictly have been one of actions but one of circumstances which for him would give rise to a remedy under the Aquilian Act. The difference from Gaius is to be explained by the fact that Justinian was prepared to provide a remedy under the act where a classical jurist would not, that is, where loss had been caused, but indirectly and not as the result of actual damage to property. For this new circumstance, in contrast to Gaius' treatment, Justinian reserves the term 'action on the case'. A further complication is that the classical jurists also give an action on the case in this instance. The explanation is that there are two different types of action on the case: one was modelled on a statutory action in the sense we mentioned above, and the other was quite independent. The classical jurists provided the latter type for this circumstance; for them it was independent of the Aquilian Act. Justinian, however, introduced it into his Aquilian title. A question is raised as to the effect of this innovation. Probably what had earlier been an independent action on the case was thereby changed into one modelled on the Aquilian Act; thus, for example, damages would now be the highest value in the previous 30 days. With the seeming abandonment of the requirement that there be damage to property, in theory the act became competent for all cases of loss, even pure economic loss, carelessly or deliberately caused. However, in practice it was not extended beyond situations like the compassionate release of a slave where, although there was no actual damage, the owner had lost a piece of physical property.

Sections one and three penalised certain positive acts of killing and wounding which were done directly. There was therefore never liability in the action on the very words of the Aquilian Act for a simple failure to act. However, once the requirement that loss be caused directly was relaxed, there were circumstances, as in J.4.3.6, in which a defendant was held liable in a policy action for his failure to act. The problem raised by these cases is to determine the criteria distinguishing them from those instances of a failure to act which did not give rise to liability. Where the jurists do give an action it is on the grounds that the defendant has been at fault, but this merely expresses their belief that he was to blame for the loss in question, and does not make the reasons for the decision clear. Modern legal systems differentiate between a pure omission, which does not give rise to liability, and an omission in circumstances in which the defendant had a duty to act, which does, but there is no evidence to suggest that the Roman jurists thought in such terms. Possibly they only gave a policy action where they regarded the failure to act simply as one way in which a positive act was done carelessly. Thus the doctor in J.4.3.6 was liable on the grounds that his act of caring for the slave was not up to standard. Similarly, the case of a person who falls asleep at a furnace,[86] which has been described as the only clear instance in which the jurists give an action for an omission,[87] can be explained as a case of the positive act of watching which was done carelessly.

Damages under section one were assessed at the highest value which the slave or animal held within the previous year (J.4.3.9). If a person killed a one-eyed slave who had been whole within the year, it was the highest market price of the slave when unimpaired which was recovered. Thus Justinian notes that the owner would sometimes receive more than his actual loss, which explains why the Aquilian actions were penal. Whereas damages functioned as a penalty because they were referred to the market value, it was nevertheless a penalty which was assessed with reference to, and therefore also intended to compensate the owner for, his loss. However, by the killing, an owner might suffer loss over and above the simple market value of the body. This provided the impetus for the inclusion, as the result of juristic interpretation (J.4.3.10), of additional forms of loss. Where a slave had been instituted heir but then killed after the death of the testator but before his owner could authorise him to accept, the value of the inheritance was included in the damages. This is an example of what is called *lucrum cessans*. Where one of a pair of animals was killed, damages included the amount by which the survivor was now worth less through no longer being part of a team. This is called *damnum emergens*.

Both these forms of loss were objective in the sense that they would have been reflected in the market price of the property. For example, the price of the slave who had been instituted heir would have included the value of the inheritance had the owner been able to authorise him to accept it. In developed classical law the approach was more subjective, in that the jurists allowed the owner to recover his personal financial interest even if this did not correspond with the market price. Thus if the slave possessed a market value of 10 but for some reason within the previous year had been worth 20 to the plaintiff, the latter was the amount recoverable (Paul, D.9.2.55). On the other hand if a slave had been killed but it could be shown that the plaintiff suffered no loss, some jurists held that he could recover nothing.[88] The function of damages by this time was primarily to compensate the owner for his personal loss, but because this was assessed with reference to the highest value of the property in the previous year they must often still have had a penal effect.

Damages under section three comprised the highest value of the object in the 30 days preceding the wrongful act. The word 'highest' does not appear in the section, so the jurists exercised a discretion and sometimes referred to the time when the value was highest and sometimes to when it was less (G.3.218). Sabinus decided, however, that the section should be interpreted as if it contained the word 'highest' on the grounds that this had been the intention of the plebeian assembly. When the act was introduced, section three was restricted to cases of serious damage for which the recovery of the highest value is not unreasonable. But what was the position once it had been interpreted to include instances of minor injury? It is argued that once this development had occurred, damages comprised not the highest value but the owner's loss, the 30-day rule no longer being applied. If this is correct, a difficulty is raised by Justinian's approval (J.4.3.15) of Sabinus' decision because, long after damages supposedly comprised mere loss, they are presented as being the highest value in the preceding 30 days (J.4.3.14). One solution is to assume that, for developed law, under section three damages were for the owner's loss but that its extent was decided with reference to the highest value the object held in the preceding 30 days, including also *lucrum cessans* and *damnum emergens*. Yet this is not what Justinian says.

There is one case, not mentioned by Justinian, in which there is no obvious way in which the 30-day rule could have been applied. The action on the very words of the Aquilian Act was always restricted to the owner of damaged property. No remedy was available to the freeman who had been injured, because he was not owner of his limbs. However, by developed classical law, under section three, a policy action was allowed to a freeman who had been carelessly injured by another, provided of course that he could satisfy the normal requirements of fault and causation. Probably the action was first available in the case of a child within authority and then extended to independent persons. Since a free person possessed no market value, compensation could be claimed only for medical expenses and loss of earnings.

(e) Contempt
(J.4.4)

As emerges from the opening paragraphs of Justinian's account, the word *iniuria* is used in both a general and a specific sense. In the general sense it means 'all wrong conduct'. More specifically, in turn it may denote the species of fault necessary for Aquilian liability, the moral or philosophical notion of injustice, or, as is relevant here, the delict of *iniuria*, where *iniuria* may be translated as 'contempt'. In its developed form, very broadly speaking, this delict comprised two quite different wrongs: deliberately inflicted physical injury and insult. What these two wrongs have in common is that each constitutes an affront to an individual, the one to his person, the other to his dignity or standing (Ulpian, D.47.10.1.2). This combination evolved through a process of development beginning, so far as recorded knowledge goes, with a group of provisions in the *Twelve Tables* on physical injury. The provisions form, not so much one single delict of physical injury, but three separate delicts each dealing with a specific type of physical assault. Though mentioned briefly in J.4.4.7 these are more fully described by Gaius (G.3.223). For a damaged limb (that is, if a limb was maimed or destroyed) or part of the body, such as an eye, ear, or finger, or perhaps even teeth removed, the penalty was the infliction of a like wound on the offender (retaliation) unless he was able to agree with the victim upon terms of compensation.[89] It is possible that compensation was most readily agreed where the injury had not been inflicted deliberately. If a bone was broken a financial penalty was imposed; 300 asses where the victim was a freeman and 150 where he was a slave. For all other contempts, trifling physical assaults of which a blow or slap in the face are examples, the penalty was 25 asses. The specific mention of a slave in the second provision creates a problem. Were slaves excluded from both the first and third provisions or were they included in the first, the penalty being retaliation exacted from a slave of the offender or payment of agreed compensation, but excluded from the third on the ground that trifling injuries to, or assaults on, slaves did not merit punishment?

Both Gaius and Justinian point, by implication, to the smallness of the financial penalties for 'a broken bone' and 'all other contempts', explaining that they are a consequence of the extreme poverty of the early Republic. Certainly an awareness that fixed pecuniary penalties established in ancient times no longer constituted adequate recompense or deterrence was an important factor in the introduction of a remedy for contempt in the praetor's edict. Such, at any rate, was the view of the early classical jurists. Aulus Gellius[90] cites from Labeo an anecdote about a rich man, Lucius Veratius, who amused himself by slapping free men in the face and then immediately paying them 25 asses. As a result, says Labeo, the praetor replaced the old rules of the *Twelve Tables* on contempt with an edict which provided that the compensation was to be set by a court of assessors. Whether or not all the details of this account are accurate, it does at least give the impression that the initial intervention of the praetor was, in Labeo's opinion, confined to the case of 'contempt', that is, it was confined to the third provision of the *Twelve Tables* and excluded provisions one and two.[91] Although there is some doubt on this point, if the more serious injuries were not included in the original edictal formulation they were certainly added later, well before the time of Gaius. On this analysis the question why, in the first instance, the praetor chose to reform the least serious case of physical assault remains unanswered.[92]

The main feature of the procedure introduced by the praetor, perhaps towards the end of the 3rd century BC,[93] was that in place of the fixed pecuniary penalty applicable to 'contempt', an action for contempt was introduced which related the assessment of compensation to the severity and circumstances of the injury. The pattern formula[94] which the praetor gave in the edict is instructive in a further respect. It puts the case in which one person hits another in the face with his fist. This is a physical assault, but one where the victim may be more concerned with the hurt to his feelings than with any possible physical injury. In other words, the praetor selects as a typical example for the operation of the new remedy a case which already contains an element of contempt. In this way a bridge is provided to the edictal clauses introduced later, which promise a remedy for various kinds of contemptuous behaviour although there has been no physical assault. The principal clauses which were added are summarised by J.4.4.1; they relate to contempt by writing or conduct. Although it is not mentioned in the *Institutes*, another important class of contempt which came to be included under the edict is that constituted by words alone.

All these clauses introduced independent praetorian wrongs which gave rise to quite separate actions on the case. However, through interpretation, by the early Principate the jurists had widened the concept of contempt, as contained in the original edict, beyond that of physical assault. The result was that the 'general edict', as it became known (Ulpian, D.47.10.15.26), absorbed all the other edicts which henceforth became superfluous, and all the wrongs now gave rise to the one action for contempt.

The wrongs as contained in the above edictal clauses constituted the main elements of contempt; however, by classical law the delict was far more wide-ranging. It comprised an almost infinite number of wrongful acts: not only physical assaults, written or oral insults, preventing a person from exercising his public or private rights, such as refusing him the use of the public baths or his seat at the theatre, but in fact any act whose common element was contempt, the essence of which was deliberate affront or insult to an individual's personality. Reflecting the possible range of the delict, unusually for an action on the case, the action for contempt contained a statement of facts alleged in which the details of the contempt had to be specified.

For the developed edictal law it is clear that 'intent' was a necessary element of the delict, in the sense that the act must permit the inference that it was designed to affront the individual. In the case of acts the essence of which was insult, there must have been an 'intention to insult' inferable from the nature of the act or the circumstances under which it was performed; in the case of acts the essence of which was the infliction of physical injury, the act must have been performed intentionally, but no separate and distinct intention to insult needed to be proved. The position in early law is not clear. It is unlikely that there was a specific requirement of intention in the sense that only deliberate acts entailed liability, yet equally there may have been no liability where the injury was accidental. Questions of liability were probably determined according to the circumstances of each case, regard also being paid to the relationship of the parties.

By Gaius' time, damages under the action for contempt were assessed in two different ways. In the normal case (where the wrong was not aggravated) the plaintiff himself put a figure on his injury, thereby expressing his opinion on the gravity of the contempt. The judge might allow this figure or reduce it. In deciding upon its adequacy he would take into account the status and respectability of the victim, even if the victim were a slave (J.4.4.7). As described in J.4.4.9 there were three ways in which the wrong might be aggravated: by conduct, place, or person. One should note that injuries which had fallen under the old law of a damaged limb and a broken bone now counted as aggravated contempts. Justinian also mentions that sometimes the location of a wound might make it aggravated, as where a person is struck in the eye. It is not clear whether this is intended as a fourth class of aggravated contempt or whether it was comprised within one of the first two classifications. Where the wrong was aggravated it was the praetor himself who determined the sum to be claimed by the plaintiff. The judge, if he found the facts proved, condemned for this amount and, in deference to the praetor, usually did not, as he might, reduce the sum (G.3.224).

In order to check the proliferation of actions based upon alleged contempt, classical law allowed a defendant who had been acquitted to recover from the unsuccessful plaintiff one-tenth of the damages originally claimed (G.4.174 - 181). Justinian, noting that this was not observed in practice, pronounced that the unsuccessful plaintiff should bear the costs of the action (J.4.16.1).

An important consequence of the shift in emphasis from physical injury to contempt was an extension in the range of possible plaintiffs arising from the one act. By virtue of the degree or kind of relationship involved, as J.4.4.2 explains, a contempt of one person, such as a child within authority, might constitute contempt to another as well, for example to the child's father. Similarly contempt to a wife was also to her husband but not vice versa. In the former case it was usual for the father to bring one action in his own name and one in the name of the child. Only exceptionally, as where the father was not present, was the child himself given the action (Ulpian, D.47.10.17.10 - 22). Where the contempt was to the wife, both she and her husband had the action. As Justinian observes, if additionally the wife was a daughter within authority there would be three actions, though in this instance, according to the principle established above, her head of the family would usually bring the action in her name. One noteworthy case not mentioned by Justinian is that contempt of a dead person was contempt of his heir, who accordingly had the action.

Slaves provided special problems and were treated separately in the edict. The praetor distinguished two grounds upon which an action might be brought by the owner in respect of contempt to his slave. The first was unreasonable or excessive flogging of the slave, or putting him to torture without the consent of the owner. The action was brought in the name of the slave and was explained in the late classical period as resting on the notion of injury to the feelings of the slave. That a slave might suffer contempt is one of the instances in which the law recognises their humanity (Ulpian, D.47.10.15.35). Damages were based upon the fact that the contempt was of the slave himself. If the owner considered that the defendant's purpose was to bring contempt on him, he might allege this as an additional feature and claim higher damages. The action lay to recover damages only for the element of contempt; with respect to loss caused to the owner through physical injury to the slave, the Aquilian action was appropriate. Similarly, where a free person brought the action for contempt he was seeking damages for the affront but where, either alternatively or additionally, he brought the policy action under the Aquilian Act the damages were based on his material loss.

The second ground on which an action might be brought for contempt to a slave was much more general. The praetor promised an action in other cases provided cause could be shown. This was interpreted by the jurists to mean that the act must constitute an aggravated contempt, as where the slave was gravely mistreated. Where the maltreatment was alleged to be a contempt of the master, the action was brought in his name; otherwise it was in the name of the slave. Ordinary contempts of a slave did not give rise to an action.[95]

Neither Gaius (G.3.222) nor Justinian (J.4.4.3) bring out the complexity of the above position. They do not distinguish clearly between the two grounds established by the edict; also, they only allow the action where the treatment of the slave amounts to contempt of the owner, which they say was possible only if the maltreatment was excessive. Justinian may additionally have required proof of an intention to show contempt to the owner, where Gaius presumed such an intention from the mere fact that the behaviour was excessive.

Possible defences to contempt are barely considered by Justinian. J.4.4.12 states the rule that the action is barred if the victim shows no anger, the rationale being that it is unjust for a victim to raise a claim for damages where he has once shown that he has passed over and condoned the contempt. The rule that the action must be brought within a year of the wrong derives from the same perspective. Thus condonation, whether explicitly or implicitly manifested, constituted a defence to the action. Omitted altogether is the defence of truth. Written or verbal statements that could be proved to be true did not amount to contempt. Lastly, it should be noted that, contrary to the rule in delictual remedies, the action was not available to the victim's heir. As a wrong, contempt did not offend any economic interest to which it was thought that the heir was entitled. Its essence was an offence to feelings, with the result that the action died with the victim.

From the late Republic a criminal as well as a civil remedy resulted from contempt. Criminal liability was instituted by the Cornelian Act on Contempts of the dictator Sulla (d. 78 BC), the provisions being summarised in J.4.4.8. Interpretation first brought all forms of deliberate physical assault within its scope, but by the time of Justinian all contempts were criminally punishable (J.4.4.10). The penalty provided by the act is unknown; nor is it clear whether a victim might recover damages under the civil action as well as prosecuting for criminal penalties. A late classical jurist states that it was more usual to take criminal than civil proceedings, and that the punishments were as follows: for slaves, flogging with whips and return to the owner; for free persons of low social standing, beating with cudgels;

and for other free persons, temporary exile or deprivation of certain rights (Hermogenian, D.47.10.45).

(f) Praetorian delicts

Neither Gaius nor Justinian mentions various special cases in which the praetor granted an action on the ground of behaviour held to be wrongful. There were three main examples of these 'praetorian delicts'. Where someone had been induced through intimidation, for example threats of death or physical injury, to enter into a transaction, he had the action on account of duress against his oppressor for fourfold damages, and after a year for simple damages. The damages were exacted only where the defendant, after the judge had found the fact of intimidation proved, failed to make good the plaintiff's loss. Where applicable, a defence based on intimidation was available. Intimidation was also a ground for restitution to the *status quo* (D.4.2).

Fraud, defined in classical law as 'every kind of cunning, trickery or contrivance practised in order to cheat, trick or deceive another' (Servius, in Ulpian, D.4.3.1.2), was the basis for the action for fraud, which lay for single damages provided there was no other remedy available to the plaintiff.[96] As in the above case, the defendant was only liable in damages where, after the judge had found the fact of fraud proved, he failed to make good the plaintiff's loss. Fraud was also a ground for restitution to the *status quo*.

The action for corruption of a slave (J.4.1.8; D.11.3) was granted for double damages against a person who deliberately made the moral, mental, or physical condition of the plaintiff's slave worse, as by inducing him to commit a delict. The damages which were doubled included not only the amount by which the slave was now worth less, but also the value of any property lost or damaged in consequence of the corruption.

XI. Obligations as though from Delict
(J.4.5)

Justinian's account is drawn from a work of Gaius called *Golden Words* (see D.44.7.5.4 - 6). There has been considerable modern discussion as to the unifying factor, if any, of the four cases which gave rise to obligations as though from delict. The difficulty of the exercise can be seen from the widely different principles of classification that have been suggested: strict or vicarious liability, liability based on non-intentional fault, on non-feasance by persons in special positions, or on a failing in a virtue proper to a particular role. Two general observations should be made. One is that neither Gaius nor Justinian states the basis of the classification, which is perhaps odd if it were something obvious like strict or vicarious liability. Secondly, Gaius and Justinian, in so far as they hint at reasons for the classification, merely point out that a case possesses a feature which distinguishes it from delictual liability, yet they imply that it is closer to delict than to contract. Beyond this there may have been no single feature of liability explaining the classification, except that all were cases of wrongdoing distinguishable from delict in that they arose only in the context of a special position, be it that of judge, occupier, shipmaster, or keeper of an inn or stable. Whether, as is often argued, the nature of the wrongdoing was the same in each case is open to doubt.

The judge is said to be liable on the ground that 'he makes a case his own'. The meaning to be given to this phrase is a matter of dispute, but the preference expressed here is that it refers to the fact that he has behaved in such a way (unspecified) as to have assumed the position of a party to the lawsuit in which he was the judge. In the early civil law the result is likely to have been that 'he incurred in relation to the plaintiff the same liability that the defendant would have incurred had he been convicted'.[97] The innovation of the edict introduced around 100 BC was that, in place of the state-law liability, it substituted the action on the case referred to in J.4.5 pr in which damages were for a 'reasonable' sum. The precise nature of the conduct for which the judge incurred this liability is also disputed. Recent studies have argued with some plausibility that the praetor, without regard to the judge's degree of fault, was originally concerned to penalise only failures to perform certain basic procedural functions required of him under the formulary system.[98] He was liable, for example, if he prejudiced the plaintiff through a failure to give a judgment, or either party by a failure to observe the limits of the condemnation. The reason why liability was limited to the above mentioned cases is explained by certain characteristics of the formulary system. The decision of the judge was absolutely final in respect of the particular dispute before him. Even where he failed to act properly, proceedings could not be restarted. Thus, because there was no system of appeals, the only recourse for the prejudiced party was against the judge himself. He was not liable for other than procedural failings because, not only were these more serious in the sense that if the judge himself abused the established procedures it went to the root of the operation of the whole system of litigation, but he was in fact chosen by the parties themselves, who therefore could not complain if in other respects he turned out to be unsatisfactory.

By the time of Justinian there had been important developments in the range of the judge's liability, but there is controversy both as to their nature and when they occurred. One could read the texts (Ulpian, D.5.1.15.1; Gaius, D.44.7.5.4; Gaius, D.50.13.6; J.4.5 pr) as showing that in Justinian's time the judge was liable not only for judgments which deliberately twisted the law, as where he was biased or corrupt, but also for wrong judgments in which the error arose from some act of carelessness on his part, or even for any wrong judgment whether he had been careless or not. The main difficulty is to determine the meaning of *per impruden-*

tiam (ignorance) which occurs in the texts attributed to Gaius (D.44.7.5.4; D.50.13.6) and which is adopted by J.4.5 pr. It is not certain that it implies liability for any judgment which turns out to be wrong, even where non-deliberate fault can be imputed to the judge.

Two further points should be made. There were a number of statutes, dating from the late Republic, which imposed criminal liability on judges for various kinds of misconduct. Before the end of the classical period, for example, a judge who received bribes was subject to punishment distinct from that imposed for making a case his own. Secondly, with the abandonment of the formulary system the possibility of appeals reduced the significance of the remedy against the judge. It would only be useful where the opportunity for a successful appeal had been lost, as where a crucial witness had died, or perhaps also where a litigant sought to recover the additional expense involved in bringing the appeal.

The action on the case concerning things thrown or poured was introduced as a public safety measure and lay against the occupier of premises for harm done by an object thrown or poured on to a place where it was usual for people to be; this included not only public roads but parks and gardens. The implication from the statement in J.4.5.1 that the occupier 'often' was liable for the fault of others is that on occasion he himself might be the one who threw the object. However, this need not be so; the utility of the remedy was that the occupier was liable merely on proof that it was from his premises that the object came. No doubt an action in delict could be brought directly against the wrongdoer where he was known, but in practice recourse will nearly always have been to this remedy, in which it was unnecessary to establish his identity. J.4.5.1 summarises the penalties established by the edict. Where a freeman was killed the action was popular, i.e. in principle it was allowed to anyone for a period of one year, though preference was given to a person with an interest in the matter or who was related to the deceased by blood or marriage (Ulpian, D.9.3.5.5).

The person who had things placed or hanging above a public way which might cause harm if they fell was also liable in a popular action. J.4.5.1, following Gaius, suggests that his liability was similar to that of the occupier with respect to what was thrown or poured from his apartment. Yet, besides the level of the penalty, there were other differences. Firstly, the action did not necessarily lie against the occupier; the edict was interpreted to mean that it lay against anyone responsible for placing or keeping placed such an object on a building (Ulpian, D.9.3.5.8). Where a householder allowed someone else to do the placing, it seems that either might be liable. Secondly, knowledge may have been a necessary condition of liability under this action; this meant knowledge of the existence and the position of the object, not of the fact that it constituted a danger.[99] Finally, it has been suggested that liability in this instance was not a true example of an obligation, let alone one arising as though from delict, since, being predicated upon the absence of damage, no actual obligation, as this was understood by the jurists, could be said to have arisen.[100] If the object did fall and cause harm the appropriate remedy was that concerning things poured or thrown.

The praetor imposed on shipmasters, innkeepers, and stablekeepers who undertook the safekeeping of goods entrusted to them a strict contractual liability in the event of their loss or damage. In addition, the praetor imposed a liability as though from delict on the same persons, who are said to be at fault in using the labour of unsatisfactory people, with respect to theft or damage of any property of a customer by one of their employees. In the case of the innkeeper, liability also extended to theft or damage committed by permanent guests. J.4.5.3 mentions only a liability for deceit and theft; deceit probably refers generally to fraudulent attempts to deprive the owner of his property. The action on the case was given for double the amount of the loss sustained by the owner.

Select Bibliography

Obligations Contracted by Conduct

Kelly, J.M., 'A Hypothesis on the Origin of *Mutuum*', *IJ* (new ser.) 5 (1970) pp. 156-63.
Litewski, W., 'Depository's Liability in Roman Law', *Archivio Giuridico* 190 (1976) pp. 3-78.
Gordon, W.M., 'Observations on *Depositum Irregulare*', *Studi in onore di A. Biscardi* III (Milan, 1983) pp. 363-72.
Zimmermann, R., *The Law of Obligations: Roman Foundations of the Civilian Tradition* (Oxford, 1996) chs 6 and 7.

Obligations by Words and their Applications

Nicholas, B., 'The Form of the Stipulation in Roman Law', *LQR* 69 (1953) pp. 63-79, 233-52.
———, 'Verbal Forms in Roman Law', *Tul L Rev* 66 (1992) pp. 1610-13.
Riccobono, S., *Stipulation and the Theory of Contract*, trans. J. Kerr Wylie and rev. B. Beinart (Amsterdam, 1957).
Zimmermann, R., *The Law of Obligations: Roman Foundations of the Civilian Tradition* (Oxford, 1996) pp. 68-75, 78-82, 114-52.

Obligations by Agreement: Sale

Mackintosh, J., *The Roman Law of Sale* (Edinburgh, 1907).
Zimmermann, R., *The Law of Obligations: Roman Foundations of the Civilian Tradition* (Oxford, 1996) chs 8, 9, and 10.
de Zulueta, F., *The Roman Law of Sale* (Oxford, 1945).

Obligations by Agreement: Hire

Frier, B.W., 'Tenant's Liability for Damage to Landlord's Property in Classical Roman Law', *ZSS* (rom. Abt.) 95 (1978) pp. 232-69.
———, *Landlords and Tenants in Imperial Rome* (Princeton, 1980).
———, 'Law, Economics, and Disasters Down on the Farm: "*Remissio Mercedis*" Revisited', *BIDR* (3rd ser.) 31-32 (1989-90) pp. 237-70.
Martin, S.D., *The Roman Jurists and the Organisation of Private Building in the Late Republic and Early Empire* (Brussels, 1989).
Munro, C.H., *Digest XIX.2. Locati Conducti* (Cambridge, 1891).
Thomas, J.A.C., 'The Nature of Merces', *AJ* (1958) pp. 191-9.
———, '*Locatio* and *Operae*', *BIDR* (3rd ser.) 3 (1961) pp. 231-47.
———, '*Non Solet Locatio Dominium Mutare*', in *Mélanges P. Meylan* I (Lausanne, 1963) pp. 339-57.
Zimmermann, R., *The Law of Obligations: Roman Foundations of the Civilian Tradition* (Oxford, 1996) chs 11 and 12.

Obligations by Agreement: Partnership

Daube, D., '*Societas* as a Consensual Contract', *CLJ* 6 (1938) pp. 381-403 (= Cohen, D., and Simon, D. (eds), *Collected Studies in Roman Law* I (Frankfurt-am-Main, 1991) pp. 37-59).
———, '*Consortium* in Roman and Hebrew Law', *Juridical Review* 62 (1950) pp. 71-91.
Munro, C.H., *Digest XVII.2 Pro Socio* (Cambridge, 1902).
Zimmermann, R., *The Law of Obligations: Roman Foundations of the Civilian Tradition* (Oxford, 1996) ch. 15.

Obligations by Agreement: Mandate

Gordon, W.M., 'The Liability of the Mandatary', *Synteleia V. Arangio-Ruiz* I (Naples, 1964) pp. 202-5.
MacCormack, G., 'The Liability of the Mandatary', *Labeo* 18 (1972) pp. 156-72.
Nörr, D., 'Reflections on Faith, Friendship, Mandate', *IJ* (new ser.) 25-27 (1990-92) pp. 302-10.
Walker, B., *Mandati vel Contra. Digest XVII.1* (Cambridge, 1879).
Zimmermann, R., *The Law of Obligations: Roman Foundations of the Civilian Tradition* (Oxford, 1996) ch. 13.

Innominate Contracts and Pacts

Buckland, W.W., '*Aestimatum*', *LQR* 43 (1927) pp. 74-80.
———, '*Aestimatum*', *LQR* 48 (1932) pp. 495-505.
MacCormack, G., 'Contractual Theory and the Innominate Contracts', *SDHI* 51 (1985) pp. 131-52.
Zimmermann, R., *The Law of Obligations: Roman Foundations of the Civilian Tradition* (Oxford, 1996) ch. 17.

5. Obligations

Obligations as though from Contract

Liebs, D., 'The History of the Roman *condictio* up to Justinian', in N. MacCormick and P. Birks (eds), *The Legal Mind: Essays for Tony Honoré* (Oxford, 1986) pp. 163-83.
Zimmermann, R., *The Law of Obligations: Roman Foundations of the Civilian Tradition* (Oxford, 1996) ch. 14.

Delict: Theft

Birks, P., 'The Case of the Filched Pedigree: D.47.2.52.20', in *Sodalitas* II (Naples, 1984) pp. 731-48.
Ibbetson, D., 'The Danger of Definition: *Contrectatio* and Appropriation', in A.D.E. Lewis and D.J. Ibbetson (eds), *The Roman Law Tradition* (Cambridge, 1994) pp. 54-72.
Jolowicz, H.F., *Digest XLVII.2. De Furtis* (Cambridge, 1940).
Kelly, J.M., *Roman Litigation* (Oxford, 1966) pp. 141, 161.
Pauw, P., 'Historical Notes on the Nature of the *Condictio Furtiva*', *SALJ* 93 (1976) pp. 395-400.
Pugsley, D., 'Furtum in the XII Tables', *IJ* (new ser.) 4 (1969) pp. 139-52.
———, '*Contrectatio*', *IJ* (new ser.) 15 (1980) pp. 341-55.
———, '*Animus Furandi*', in *Sodalitas* V (Naples, 1984) pp. 2419-26.
Stein, P., 'School Attitudes in the Law of Delicts', in *Studi in onore di A. Biscardi* II (Milan, 1982) pp. 281-7.
Watson, A., 'D.47.2.52.20: The Jackass, the Mares and "*Furtum*" ', in *Studi in onore di E. Volterra* II (Milan, 1971) pp. 445-9.

Delict: The Aquilian Act

Andrews, N.H., '*Occidere* and the *Lex Aquilia*', *CLJ* 46 (1987) pp. 315-29.
Birks, P., 'Other Men's Meat: Aquilian Liability for Proper User', *IJ* (new ser.) 16 (1981) pp. 141-85.
———, 'Cooking the Meat: Aquilian Liability for Hearths and Ovens', *IJ* (new ser.) 20 (1985) pp. 352-77.
———, 'Doing and Causing to be Done', in A.D.E. Lewis and D.J. Ibbetson (eds), *The Roman Law Tradition* (Cambridge, 1994) pp. 38-46.
Lawson, F.H., and Markesinis, B.S., *Tortious Liability for Unintentional Harm in the Common Law and the Civil Law* I, II (Cambridge, 1982) (contains references to earlier literature).
MacCormack, G., 'Juristic Interpretation of the *Lex Aquilia*', *Studi in onore di C. Sanfilippo* I (Milan, 1982) pp. 255-83.
Nörr, D., '*Causam Mortis Praestare*', in N. MacCormick and P. Birks (eds), *The Legal Mind: Essays for Tony Honoré* (Oxford, 1986) pp. 203-17.
Robaye, R., 'Remarques sur le concept de faute dans l'interprétation classique de la *lex Aquilia*', *RIDA* (3rd ser.) 38 (1991) pp. 333-84.
Zimmermann, R., *The Law of Obligations: Roman Foundations of the Civilian Tradition* (Oxford, 1996) chs 29 and 30.

Delict: Contempt

Birks, P., 'Lucius Veratius and the *Lex Aebutia*', in A. Watson (ed), *Daube Noster* (Edinburgh, 1974) pp. 39-51.

Daube, D., '*Ne quid infamandi causa fiat*. The Roman Law of Defamation', *Atti del congresso Verona* 3 (Milan, 1948) pp. 413-50 (= D. Cohen and D. Simon (eds), *Collected Studies in Roman Law* I (Frankfurt-am-Main, 1991) pp. 465-500).

Halpin, A.K.W., 'The Usage of *Iniuria* in the Twelve Tables', *IJ* (new ser.) 11 (1976) pp. 344-54.

Polay, E., *Iniuria Types in Roman Law* (Budapest, 1986).

Watson, A., *The Roman Law of Obligations in the Later Roman Republic* (Oxford, 1965) pp. 248-55.

———, 'Personal Injuries in the XII Tables', *TvR* 43 (1975) pp. 213-22.

Zimmermann, R., *The Law of Obligations: Roman Foundations of the Civilian Tradition* (Oxford, 1996) ch. 31.

Obligations as though from Delict

Birks, P., 'The Problem of Quasi-Delict', *Current Legal Problems* 22 (1969) pp. 164-80.

———, 'Obligations: One Tier or Two?', in Stein and Lewis (eds), *Studies in Memory of Thomas* pp. 18-38.

———, 'A New Argument for a Narrow View of *Litem Suam Facere*', *TvR* 52 (1984) pp. 373-87.

Kelly, J.M., *Roman Litigation* (Oxford, 1966) pp. 102-17.

Gordon, W.M., 'The Roman Class of Quasi-Delicts', *Temis* 21 (1967) pp. 303-10.

Pugsley, D., '*Litem Suam Facere*', *IJ* (new ser.) 4 (1969) pp. 351-5.

Stein, P., 'The Nature of Quasi-Delictal Obligations in Roman Law', *RIDA* (3rd ser.) 5 (1958) pp. 563-70.

Notes

1. See Nicholas, *Introduction* pp. 168-71.

2. See R. Evans-Jones, 'The Penal Characteristics of the *Actio Depositi in Factum*', *SDHI* 52 (1986) pp. 105-18.

3. See R. Zimmermann, *The Law of Obligations: Roman Foundations of the Civilian Tradition* (Oxford, 1996) pp. 181ff.

4. See D. Daube, 'Did Macedo Murder his Father?', *ZSS* (rom. Abt.) 65 (1947) pp. 261-311.

5. There is a full discussion in section X(b).

6. See section X(b).

7. *Collatio legum Mosaicarum et Romanarum* 10.7.11.

8. See G. MacCormack, 'Gift, Debt, Obligation and the Real Contracts', *Labeo* 31 (1985) pp. 131-54.

9. For Justinian, cf. C.4.34.11.

10. J.A.C. Thomas, *Textbook of Roman Law* (Amsterdam, 1976) p. 278.

11. G. MacCormack, 'The Oral and Written Stipulation in the Institutes', in Stein and Lewis (eds), *Studies in Memory of Thomas* pp. 96-108.

12. Cf. J.3.20.8 on guarantors and C.8.37.1 of AD 208.

13. See Buckland, *Textbook* p. 441.

14. See Zimmermann, *The Law of Obligations* pp. 95ff.

15. See also Buckland, *Textbook* p. 443.

16. A full discussion of the history of guarantors is found in J.3.20.

17. Buckland, *Textbook* p. 453.

18. W. Litewski, ' "*Litis contestatio*" et obligations solidaires passives dans les "*bonae fidei iudicia*" en droit romain classique', *RHDFE* 54, (1976) pp. 149-75.

19. See Buckland, *Textbook* p. 426.

20. The position of sons within authority and slaves was different; see J.3.19.6; Buckland, *Textbook* p. 438.

21. Note the similarities in form with adstipulation discussed in J.3.16.

22. *Nov*.4 (AD 535).

23. See also Theophilus, *Paraphrase* 3.21.

24. See Buckland, *Textbook* p. 460; cf. Thomas, *Textbook* p. 267, who argues that the two sorts of entry were distinguishable by the fact that the cash entries recorded the ground of the receipt of payment, e.g. 'Advanced to X 100 by way of loan/purchase'.

25. For the classical controversy on the use of the contract by foreigners see G.3.132, 133.

26. It is sometimes held that the plea that the money had not been paid was available as a defence to an action alleging a stipulation, as where a document stated, 'A promised to pay B 100'. Cf. Buckland, *Textbook* p. 442; Thomas, *Textbook* pp. 268-9; Zimmermann, *The Law of Obligations* pp. 93-4.

27. The version in the *Institutes* is not quite the same with respect to (1), since it distinguishes between the case where the documents have been written by the parties themselves and that where they have been written out by someone else. In the latter case the documents do not constitute a fully binding contract of sale unless they have been signed by the parties, signature not being necessary in the former case.

28. Amongst others: A. Watson, '*Arra* in the Law of Justinian', *RIDA* (3rd ser.) 6 (1959) pp. 385-9; A.M. Honoré, '*Arra* as You Were', *LQR* 77 (1961) pp. 172-5; T.H. Tylor, 'Writing and *Arra* in Sale under the *Corpus Iuris*', *LQR* 77 (1961) pp. 77-82; J.M. Thomson, 'Arra in Sale in Justinian's Law', *IJ* (new ser.) 5 (1970) pp. 179-87; M.C. Marasinghe, '*Arra* – Not in Dispute', *RIDA* (3rd ser.) 20 (1973) pp. 349-53.

29. J.A.C. Thomas, '*Arra* in Sale in Justinian's Law', *TvR* 21 (1956) pp. 253-78.

30. B. Frier, 'Roman Law and the Wine Trade', *ZSS* (rom. Abt.) 100 (1983) p. 263.

31. J.A.C. Thomas, 'Venditio Hereditatis and Emptio Spei', *Tul L Rev* 33 (1958-59) pp. 541-50; D. Paling, '*Emptio Spei* and *Emptio Rei Speratae*', *IJ* (new ser.) 8 (1973) pp. 178-82. See also D. Daube, 'Purchase of a Prospective Haul', in *Studi in onore di U.E. Paoli* (Florence, 1955) pp. 203-9.

32. See also P. Stein, *Fault in the Formation of Contract in Roman Law and Scots Law* (Edinburgh, 1958) pp. 61-83; J.A.C. Thomas, 'The Sale of *res extra commercium*', *Current Legal Problems* 29 (1976) pp. 136-49; R. Evans-Jones, 'The Origins of Justinian's Institutes 3.23.5', *CLJ* 53 (1994) p. 473; R. Evans-Jones and G. MacCormack, 'The Sale of *res extra commercium* in Roman Law', *ZSS* (rom. Abt.) 112 (1995) pp. 330-51.

33. See D. Daube, 'Generalisations in D.18.1, *de contrahende emptione*', in *Studi in onore di V. Arangio-Ruiz* I (Naples, 1952) pp. 185-200.

34. D. Daube, 'Certainty of Price', in D. Daube (ed), *Studies in the Roman Law of Sale* (Oxford, 1959) pp. 26-45.

35. Cf. id. pp. 21-6.

36. J.A.C. Thomas, 'Marginalia on *certum pretium*', *TvR* 35 (1967) pp. 77-89; cf. Daube, 'Certainty of Price', pp. 9-20.

37. See C.4.44.2 (Diocletian and Maximian), C.4.44.8 (same), both considered interpolated. Also, H.F. Jolowicz, 'The Origin of *laesio enormis*', *JR* 49 (1937) pp. 50-72; H.T. Klami, '*Laesio enormis* in Roman Law', *Labeo* 33 (1987) pp. 48-63.

38. See Frier, 'Roman Law and the Wine Trade', p. 257.

39. Cf. Thomas, *Textbook* pp. 231-2.

40. Possibly no distinction was made in this context between the innocent and fraudulent seller; see Stein, *Fault in the Formation of Contract* p. 52.

41. See G. MacCormack, 'Alfenus Varus and the Law of Risk in Sale', *LQR* 101 (1985) pp. 573-86.

42. Paul's *Sentences* 2.17.4. A special remedy lay where the seller in mancipating land misstated the acreage. The buyer by the *actio de modo agri* might recover double the value of the difference between the acreage mancipated and that declared to be mancipated. By the end of the classical period it seems that the same remedy was available under the contractual action, though this was perhaps confined to the case where the seller had deliberately lied.

43. Cicero, *De Officiis* 3.65.

44. See A.M. Honoré, 'The History of the Aedilitian Actions from Roman to Roman-Dutch Law', in D. Daube (ed), *Studies in the Roman Law of Sale* (Oxford, 1959) pp. 132-59.

45. Cf. B. Nicholas, '*Dicta Promissave*', in D. Daube (ed), *Studies in the Roman Law of Sale* (Oxford, 1959) pp. 91-101.

46. Where Julian is cited as going even further.

47. See R. Powell, 'Eviction in Roman Law and English Law', in D. Daube (ed), *Studies in the Roman Law of Sale* (Oxford, 1959) pp. 78-90.

48. See A. Lewis, 'The Trichotomy in *Locatio Conductio*', *IJ* (new ser.) 8 (1973) pp. 164-77.

49. See A.M. Pritchard, 'Sale and Hire', in D. Daube (ed), *Studies in the Roman Law of Sale* (Oxford, 1959) pp. 1-8.

50. Buckland, *Textbook* pp. 506-7.

51. See id. p. 519.

52. There were some qualifications to this rule, possibly only in later law; see id. p. 512.

53. A. Watson, *Contract of Mandate in Roman Law* (Oxford, 1961) pp. 61-77.

54. See W.M. Gordon, 'Agency and Roman Law', in *Studi in onore di C. Sanfilippo* III (1983) pp. 341-9.

55. The limits within which this action was given are not clear; see Buckland, *Textbook* p. 519, who suggests that the innovation was that the action was given even when the mandate was only for an isolated transaction, the mandatary in that case therefore not being an *institor*.

56. Cf. Julian, D.14.3.12; Paul, D.46.5.5, which both mention only the *institor* arrangement.

57. Prior to the introduction of the praetorian action on the case, the only remedies for the party who had performed was the action for fraud in appropriate circumstances, or a *condictio* to recover any property handed over in return for the 'consideration' that failed.

58. On the relationship between the pact of *receptum* and the quasi-delictual liability of shipmasters, innkeepers, and stablekeepers, see Zimmermann, *The Law of Obligations* p. 517.

59. See J.A.C. Thomas, *The Institutes of Justinian* (Amsterdam, 1975) p. 251.

60. There was some relaxation of this position in the time of Justinian; see Buckland, *Textbook* p. 569.

61. Id. p. 568.

62. E. Levy, *Privatstrafe und Schadensersatz* (Berlin, 1915).

63. See H. Ankum, 'Actions by which We Claim a Thing (*res*) and a Penalty (*poena*) in Classical Roman Law', *BIDR* (3rd ser.) 24 (1982) pp. 15-39.

64. See A. Watson, 'The Definition of *Furtum* and the Trichotomy', *TvR* 28 (1960) pp. 197-210.

65. P. Birks, 'A Note on the Development of *Furtum*', *IJ* (new ser.) 8 (1973) pp. 349-55.

66. G.D. MacCormack, 'Definitions: *Furtum* and *Contrectatio*', *AJ* (1979) pp. 129-47.

67. W.W. Buckland, '*Contrectatio*', *LQR* 57 (1941) pp. 467-74; A. Watson, '*Contrectatio* as an Essential of *Furtum*', *LQR* 77 (1961) pp. 526-32; J.A.C. Thomas, 'Contrectatio, Complicity and Furtum', *IURA* 13 (1962) pp. 70-88; B. Nicholas, 'Theophilus and *Contrectatio*', in Stein and Lewis (eds), *Studies in Memory of Thomas* pp. 118-24.

68. J.A.C. Thomas, '*Animus Furandi*', *IURA* 19 (1968) pp. 1-32.

69. See G. MacCormack, '*Visus vel Deprehensus*: Inst. 4.1.4', in R. Feenstra et al. (eds), *Collatio Iuris Romani. Études Dediées à H. Ankum* I (Amsterdam, 1995) pp. 275-83.

70. H.F. Jolowicz, *Digest XLVII.2 De Furtis* (Cambridge, 1940) p. lxix.

71. J.G. Wolf, '*Lanx* und *Licium*. Das Ritual der Haussuchung im altrömischen Recht', in *Sympotica Franz Wieacker* (Göttingen, 1968) pp. 59-79.

72. D. Daube, *Studies in Biblical Law* (New York, 1969) pp. 259-305.

73. Id. p. 282

74. G.D. MacCormack, '*Ope Consilio Furtum Factum*', *TvR* 51 (1983) pp. 271-93.

75. Cicero, *Pro Tullio* 8; M. Kaser, *RPR* I p. 627.

76. J.4.6.19; see H. Ankum, 'Gaius, Theophilus and Tribonian and the *Actiones Mixtae*', in Stein and Lewis (eds), *Studies in Memory of Thomas* pp. 4-17.

77. The matter is not entirely clear; cf. Ulpian, D.47.8.2.24.

78. See D. Daube, 'On the Third Chapter of the *Lex Aquilia*', *LQR* 52 (1936) pp. 253-68.

79. J.M. Kelly, 'The meaning of the *Lex Aquilia*', *LQR* 80 (1964) pp. 73-83.

80. Id.; D. Pugsley, *Roman Law of Property and Obligations* (Cape Town, 1974) pp. 101-10.

81. H.F. Jolowicz, 'The Original Scope of the *Lex Aquilia* and the Question of Damages', *LQR* 38 (1922) pp. 220-30.

82. Daube, 'On the Third Chapter', loc. cit. above note 78.

83. G. MacCormack, 'On the Third Chapter of the *Lex Aquilia*', *IJ* (new ser.) 5 (1970) pp. 164-78.

84. B. Beinart, 'The Relationship of *Iniuria* and *Culpa* in the *Lex Aquilia*', in *Studi in onore di V. Arangio-Ruiz* I (Naples, 1952) pp. 279-303.

85. G. MacCormack, 'Aquilian *Culpa*', in A. Watson (ed), *Daube Noster* (Edinburgh, 1974) pp. 201-24.

86. *Collatio legum Mosaicarum et Romanarum* 12.7.7.

87. F.H. Lawson, *Negligence in the Civil Law* (Oxford, 1950; rev. 1955) p. 26.

88. See A. Rodger, 'Damages for the Loss of an Inheritance', in A. Watson (ed), *Daube Noster* (Edinburgh, 1974) pp. 289-99.

89. The meaning given to 'a damaged limb' here is not shared by all modern scholars; see P. Birks, 'The Early History of *Iniuria*', *TvR* 37 (1969) pp. 163-208.

90. *Noctes Atticae* 20.1.12, 13.

91. See D. Daube, '*Nocere* and *Noxa*', *CLJ* 7 (1939-41) pp. 23-55.

92. But cf. Birks, 'The Early History of *Iniuria*'.

93. See A. Watson, *Law Making in the Later Roman Republic* (Oxford, 1974) p. 112.

94. Lenel, *EP* pp. 397-403.

95. See W.W. Buckland, *The Roman Law of Slavery* (Cambridge, 1970) pp. 79-82.

96. A. Watson, '*Actio de Dolo* and *Actiones in Factum*', *ZSS* (rom. Abt.) 78 (1961) pp. 392-401.

97. G. MacCormack, 'The Liability of the Judge in the Republic and Principate', in *ANRW* XIV pp. 5-6.

98. D.N. MacCormick, '*Iudex Qui Litem Suam Facit*', *AJ* (1977) pp. 149-65; MacCormack, 'The Liability of the Judge'.

99. This point is a matter of dispute; cf. A. Watson, 'Liability in the *Actio de Positis ac Suspensis*', in *Mélanges P. Meylan* I (Lausanne, 1963) pp. 379-82; W.M. Gordon, 'The *Actio de Posito* Reconsidered', in Stein and Lewis (eds), *Studies in Memory of Thomas* pp. 45-55.

100. MacCormick, '*Iudex Qui Litem Suam Facit*' p. 161.

6. Actions

Ernest Metzger

The law of actions is the last of the three subjects in the institutional scheme, and is quite different from the subjects that precede it. While the earlier subjects are concerned with substantive rules and principles, the law of actions is concerned with redress. It is not quite the same as 'the law of procedure', however; it often includes matters that might easily have been treated under the law of persons or of things. This is because actions evolved as a subject over a time when procedure was not distinct from substantive law.

I. 'Action'

(J.4.6 pr; 4.15)

The word 'action' is awkward to define, not because its meaning is hard to understand from any one example, but because it appears in many different contexts and therefore evades any one definition. In Latin the word is *actio*, from the verb *agere*, which for our purposes is best translated broadly: 'to urge'. Generally, to have an action means that a person is entitled to pursue a remedy for some injustice done to him. If, for example, a person has been a victim of fraud, he might be allowed an *actio de dolo*, an 'action for fraud'. This would entitle him to go before a judge, and to urge the judge to give him relief.

From this example it might seem that an action is in fact a 'right', and that when we say a person is entitled to pursue a remedy we are saying that he has a right to relief. But to equate an action with a right is to substitute a concept that we appreciate easily, but that the Romans came to appreciate only over time. We have no difficulty in understanding that a person under certain circumstances (for example, when he is injured) has a right, that that right has an existence in the abstract, and that it is the function of the judicial machinery – the judges, tribunals, rules of procedure and evidence – to transform that right into a remedy. But in a system where the substantive law often speaks in the language of procedure ('If it appears that X ought to give 10 to Y, the judge shall condemn for 10'), and where procedure is not seen as something necessarily separate from the rest of the law, this sort of abstraction does not come easily.[1] Under such a system, someone who has suffered a wrong will probably see himself, not as a person with some abstract entitlement to be made whole, but as a person who is entitled to clear a procedural hurdle. This is made very plain in the classical definition of *actio* (Celsus, D.44.7.51) which Justinian uses at J.4.6 pr, a definition that has certain shortcomings but otherwise conveys well the limited notion of an action: 'An action is nothing but a right to go to court to get one's due.'[2] This statement shows clearly that an action, if it is a right at all, is more a right to proceed than a right to prevail, and that any definition of 'action' ought to put process at the fore. Accordingly, *actio* is often translated as 'claim' or (circuitously) as 'right of action', to show that a person, on presenting certain facts, will be allowed to follow a certain procedural agenda appropriate to those facts. That person hopes, but is not assured, that this agenda will culminate in the granting of relief. In English the word 'warrant' also has something of *actio* in this sense.

For our purposes it is satisfactory to define an action as a claim or a warrant. At the same time it is worth mentioning that *actio* appears in contexts in which the meaning is either more broad or more refined, and where 'claim' or 'warrant' will not work.

(1) To describe an action as a claim or warrant gives the impression that an action requires the intervention of the state. Yet the word *actio* may describe a purely private event, as for example in D.48.1.7 (Macer), where *actio* is used to describe simply the act of a person, an act which (in this context) brings disgrace upon that person if it becomes the subject of certain proceedings. An important example of *actio* as a private event is self-help. An *actio* may describe not only the pursuit of a remedy with the sanction of the state, but also the act of an individual privately vindicating a wrong done to him. In the earlier forms of Roman procedure, for example, there existed actions for seizing a person or his property – the *legis actio per manus iniectionem* and *legis actio per pignoris capionem* (G.4.21 - 29) – that a person might resort to without prior litigation.

Self-help is not a particularly significant institution in the developed law of actions, but is nevertheless important for illustrating the essence of the word 'action'. Roman law did not draw a firm line between a person's real and formal claim, that is, between the claim a person possesses simply by virtue of being wronged, and the claim recognised by a court.[3] Thus a person might possess a claim for redress even if the judicial machinery had not yet granted it to him.

(2) In their treatments of actions, both Gaius and Justinian discuss a great deal more than simply 'claims', and for this reason we often understand the idea of 'actions' quite broadly. Gaius, for example, devotes a fair amount of space to procedure, and even discusses a form of procedure that was all but unused in his own time. Justinian, though he does not speak directly about the earlier forms of procedure (which were obsolete by his time), does speak about more general matters of litigation, for example, pleading, interdicts, and judges. And in our other sources we find definitions of 'actio' that are very broad (see e.g. Ulpian, D.44.7.37), generally encompassing the various ways of presenting issues to a tribunal. For

these reasons 'actions' is sometimes treated as synonymous with 'procedure' or 'remedies'.

The two instances of *actio* just discussed – 'private act' and 'procedure' – are cited in order to show that 'action' may sometimes mean something more general than 'claim'. In other contexts it is used in exactly the opposite way, to indicate not simply a claim, but a particular kind of claim:

(3) An action is sometimes distinguished from an interdict. 'Interdict' describes several decrees issued by magistrates and often used to confer or protect possession. They were highly procedural in character; it would be difficult to discuss the 'substantive law of interdicts', because an interdict was a specific order granted on the presentation of specific facts. What distinguishes an interdict from an action (as described here) is principally the fact that an interdict, in form, is virtually a remedy in itself. The magistrate does not, as in an ordinary action, summarise the dispute and pass it on to someone else for resolution, but instead issues a decree himself. The decree, it is true, might not be the end of the matter, instead often requiring the resolution of a judge, or serving as a predicate for a subsequent action. But the interdict alone, as a decree issued directly by the magistrate, was administrative in character, and was therefore distinguishable from an action.

Justinian offers several different ways to classify interdicts (J.4.15); the possessory interdicts and their three divisions are worth mentioning briefly. These interdicts have in common the feature that each culminates in the grant of possession in favour of a party. The first possessory interdict is for obtaining possession, and Justinian's example is an interdict that arises in the context of succession law (J.4.15.3). This interdict existed as a part of the praetor's efforts to alter the scheme of succession under the state law with a scheme (*bonorum possessio*) based to a greater extent on blood relationship (see Chapter 4, section III(c) above). A person favoured by the praetor's innovations who wished to obtain possession of the tangible assets of the estate could request from the praetor an 'interdict *quorum bonorum*'. The interdict took the form of an order to the person in possession – in this case a person without title or who claimed to possess as heir – directing that person to restore the property to the other. Even though the interdict determined the question of possession and nothing else, this would often be the end of the dispute: efforts to establish ownership of the property against the new possessor (if the grant of *bonorum possessio* were of a certain character) would be fruitless.

The situation was different under the second division of the possessory interdicts, those for retaining possession. Here the interdicts often served only as a procedural step in advance of a proper action for ownership, and not as a final remedy. In other words, these interdicts were very much part of a pretrial tactic, made necessary by the characteristics of the ownership action.

In a perfect system of litigation, we would expect a tribunal to listen equally to competing claims for ownership and give judgment in favour of the better claim. But in the Roman system, as in modern systems, a plaintiff carries the burden of proof, and proving ownership is not an easy matter, particularly where no system of public registration exists. This means that if a person had the freedom to do so he would choose to be a defendant and not a plaintiff, leaving it to his opponent to prove ownership. The possessory interdicts, to a certain extent, allowed such a choice: if a party could put himself in possession of the disputed property, it was then left to his opponent to bring suit and try to establish ownership. Thus the battle for ownership might be preceded by a battle for possession (J.4.15.4).

In the case of immoveables, the battle for possession might begin with the pronouncement of an interdict *uti possidetis*. This would consist of a decree addressed to both parties, stating that force could not be used to dispossess whichever of the parties was innocently in possession of the property. After a complex course of proceedings, the party with the better claim under the words of the interdict would be awarded possession. This would put him in the more enviable position of defending rather than establishing ownership, if his opponent sought to claim ownership in a further proceeding. In Justinian's time the interdict *utrubi*, applying to moveables, operated in the same way, although in the classical law the phrasing of the interdict was slightly different; it allowed an innocent possessor not only to retain possession, but to recover possession that was recently lost (J.4.15.4a).

The third kind of possessory interdict is for recovering possession. The *Institutes* gives the example of a person dispossessed of immoveable property by force (J.4.15.6). The interdict described there – the interdict *unde vi* – existed as two different interdicts in the classical law but was made uniform by Justinian's time. The classical interdict *unde vi* was an order to restore immoveable property to an innocent possessor who had been evicted by force within the previous year. The classical interdict *unde vi armata* was directed against one who had dispossessed not only by force, but by armed men as well, and given the gravity of the act it was unnecessary to show that the dispossessed person had come by his possession innocently. In Justinian's synthesis of these two interdicts, violent dispossession was enough disliked that an ejector could no longer challenge the innocence of the ejectee's possession in any event.

The taxonomy that distinguishes an action from an interdict is far from perfect. In the *Institutes* Justinian (borrowing from Gaius) first says that '*All* our law is about persons, things, or actions' (J.1.2.12), and later says 'We look next at interdicts *or the actions used instead of them*' (J.4.15 pr). It is important not to worry too much over this sort of inconsistency, but simply to recognise the basic difference between an action (a claim) and an interdict (an administrative remedy).

(4) There are other so-called 'praetorian remedies' that are often distin-

guished from actions. Like interdicts, these remedies were issued by the praetor himself, sometimes after a short inquiry, and sometimes after hearing only one party. There were various forms. For example, the praetor sometimes circumvented the law and restored parties to their previous positions, put a party in possession of the other party's property as security, or exacted a promise from one party in favour of the other. Some of these remedies appear sporadically in the *Institutes*.[4]

(5) The three words *actio petitio persecutio* appear together in some contexts and are understood to refer to specific types of suits. In this narrow sense an *actio* refers to a personal action, *petitio* refers to a real action, and *persecutio* refers to a restorative action (according to Papinian, D.44.7.28), or to an extraordinary proceeding (according to Ulpian, D.50.16.178.2). The most familiar occurrence of these three words is in the so-called Aquilian stipulation, described at J.3.29.2 (see Chapter 5, section IX above). They also appear in some legislation (see the *lex Coloniae Genetivae Iuliae*, chs 125-132, *passim*).

II. The Formulary Procedure

From the 2nd century BC until the 3rd century AD, most Roman litigation was conducted according to what is called the 'formulary procedure'. The subject is omitted from the *Institutes* but discussed at length by Gaius. It is difficult to give any account of actions without at least a sketch of the formulary procedure; the subject of actions owes a great deal to its influence. Of course 'claims' existed long before it was created and continued to exist long after it was abolished. But actions as Gaius and Justinian present them are not simply lists of claims. They are claims that are classified with great acuteness and expressed with great technical precision. Their classifications are due partly to the formulary procedure, and their expression is due almost entirely to it.

Roman litigation was conducted in two phases. The first phase was public, conducted before a magistrate of the state, the praetor or aedile, charged with administering justice. The second phase was private, conducted before a judge, a private individual who need not have been a lawyer. The public phase was very brief; the magistrate would simply determine whether the litigants should be allowed to proceed and, if so, what form their action should take. The private phase was the trial itself.

The magistrate needed a scheme for determining which claims would be allowed to go forward. His duties would have become impossible if he had had to consult treatises and legislation and make a fresh decision on the suitability of every claim. Accordingly he maintained and put on display a long list announcing his intentions and expectations regarding the lawsuits he would allow. This list, the edict, contained individual entries describing actions which he was willing to grant. If a litigant came before him and requested one of the actions, the magistrate would ordinarily grant it (though he might deny it under certain circumstances, for example on account of *res judicata*). If the litigant's circumstances did not match any of the entries, he might persuade the magistrate to invent a new claim and allow it to go before a judge. If the magistrate saw fit, he might even incorporate the new claim in the edict for future cases.

The passing of the suit from the magistrate to the judge was an act that required great care. The principal problem was the judge's peculiar standing within the legal system. The judge did not hold office, but was appointed for service in a single case, and selected personally by the parties if possible. He had no special qualifications other than his wealth. He was simply a private individual who conducted the trial without even intermittent guidance from the state. The consequences were that (1) he required detailed written instructions at the outset, and (2) what he did with those instructions – his conduct of the trial, his judgment – was of no enduring importance whatsoever to the legal system.

This meant that the final, formal act of the state, the final expression of the law in a given case, was the set of instructions that the magistrate gave to the judge. These instructions, from one perspective, were the parties' pleadings, as they contained their allegations, the matters they hoped to prove. But because the allegations had to satisfy the requirements of the law as determined by the magistrate, they came into the judge's hands in a technical form, a form that permitted relief under the law. This makes the instructions the single most important item in the lawsuit, far more important than the judgment. A judgment declaring who won or lost could have little value compared to these instructions which, under this public/private system of litigation, necessarily recited for the judge's benefit the kernel of the dispute: what a party had to show in order to win.

The instructions were prepared according to formulae, composed of 'specially prepared phrases' (G.4.30). Each formula was divided into parts, and each part had a particular function. Very few actual formulae survive; one of the few that does survive is below. It was found near Pompeii and dates from the first century AD:[5]

> C. Blossius Celadus shall be the judge. If it appears that C. Marcius Saturninus ought to give 18,000 sesterces to C. Sulpicius Cinnamus, which is the matter in dispute, C. Blossius Celadus, the judge, shall condemn C. Marcius Saturninus for 18,000 sesterces in favour of C. Sulpicius Cinnamus; otherwise he shall absolve.

This formula describes an action called a *condictio certae pecuniae*, a personal claim for a particular sum of money. It was the appropriate action in cases where, among other things, a person had given a stipulation to pay money to another. We are able to classify formulae in this way and pair certain formulae with certain actions because each different formulae was drafted with a significant form of words. One part of the formula, the *intentio* (or 'principal pleading'), is particularly revealing in this respect. The *intentio* in the formula above is the phrase 'If it appears that C. Marcius Saturninus ought to give 18,000 sesterces to C. Sulpicius Cinnamus ....' The word 'if' tells us the claim is for a certain sum (quite apart from the appearance of the sum itself), and the word 'ought' tells us it is an action for a debt. If the words are altered, the action is altered. If the sum promised by the stipulation were uncertain, the formula would not order the judge to condemn 18,000 'if Cinnamus owes 18,000', but to condemn 'whatever Cinnamus owes'. The 'whatever' in the *intentio* indicates a claim for an uncertain sum. And the altered formula produces a different action, the *actio incerti ex stipulatu*.

Nearly every action is associated with a unique formula.[6] This means that one usually describes, classifies, and analyses different actions by addressing the formulae that describe each action.

III. The Law of Actions

It is not easy to understand how actions could constitute a subject by themselves. Under the heading 'law of actions' one might expect to find a list of actions with a description of each action, in the form 'if X injures Y in such a way, X will compensate Y to such a degree.' But this sort of list would be useful only if all the law were expressed in the form of individual actions. That is, in a legal system such as the *Institutes* describe, where actions constitute only one subdivision of the law, we would not expect the law of actions to be a list of every action. In such a system there is no reason to discuss, for example, the depositor's action, the ward's action, or the vindicatory action as subjects apart from deposit, guardianship, or ownership. Therefore the law of actions – at least as it appears in the institutes of Gaius and Justinian – cannot be a list of actions, but must exist somehow as a subject apart from the underlying substantive rules.

As a subject 'actions' never stayed the same for very long, but changed as its relationship with the substantive rules changed. For this reason the subject has a very different character in different historical periods. In the beginning, actions very possibly encompassed most of the law. By Justinian's time, actions in the classical sense of 'claim' were so reduced in importance that much of what appears on the subject in the *Institutes* must be read as (1) a historical description of how matters were pleaded centuries earlier, or (2) something like a discussion of rights, in the modern sense.

The customary view of actions in the earliest law was expressed famously by Henry Maine (1822-88). He does not refer directly to Rome at the time of the *Twelve Tables*, although he seems to have had it in mind, among other examples:[7]

> The primary distinction between the early and rude, and the modern and refined, classifications of legal rules, is that the Rules relating to Actions, to pleading and procedure, fall into a subordinate place and become, as Bentham called them, Adjective Law. So far as this the Roman Institutional writers had advanced, since they put the Law of Actions into the third and last compartment of their system. Nobody should know better than an Englishman that this is not an arrangement which easily and spontaneously suggests itself to the mind. So great is the ascendancy of the Law of Actions in the infancy of Courts of Justice, that substantive law has at first the look of being gradually secreted in the interstices of procedure; and the early lawyer can only see the law through the envelope of its technical forms.

Maine writes in a 19th-century style that produces memorable language, but at the expense of authorities and accuracy. We can nevertheless take his general point: rules in earlier Roman law were often expressed according to the remedial steps to be followed. The *Twelve Tables* have many examples of rules expressed in this way, e.g.,

Table 1, 14. If he has broken the bone of a free man, let the penalty be 300. If of a slave, 150.

Table 8, 11. Anyone who allows himself to be a witness or serve as a holder of the balances and then does not stand by his evidence will be untrustworthy and incompetent as a witness.

Table 11, 2. If a slave commits theft or causes damage, [he shall be given noxally].

Of course not all of the rules in the *Twelve Tables* are expressed in this way, and there are even good arguments to the effect that the *Twelve Tables* is nothing like what Maine describes.[8] But to whatever extent the *Twelve Tables* is dominated by procedure and remedies, it is clear this is not an effective way to present the law. The main criticism is that this sort of presentation is unreflective, that it leaps immediately to the question 'how much?' without stopping long to consider 'who is the wrongdoer?' or 'what is the wrong?'. A more reflective method of presenting laws would group similar rules together and consider in mass the substantive components – the 'whos' and the 'whats' – of each rule. This would allow a person better to classify a given set of facts as a particular kind of legal event. It is therefore something to be admired when, for example, jurists and legislators take up the events described under the first and third examples above, consider those events separately from any particular remedy, and then discuss them under the common rubric of delict. And that is Maine's point: that with time, the underlying substantive ideas were more frequently discussed and expressed as subjects apart from litigation, thus producing a more refined method of classification.

The law of actions is very much a function of this gradual division of substantive law from procedure. If we imagine a system where the entire law is expressed in the form of actions, 'the law of actions' would be synonymous with 'the law'. As the substantive law underlying the actions is gradually set apart for separate discussion, what remains behind is a residue of procedure, remedies, and as yet undisentangled substantive law. For lack of a better term we might call this the 'law of actions'. How closely this model in fact describes the development of Roman law is a matter of debate. Some, notably Watson, insist that Roman law scrupulously divided substantive law from procedure, even as early as the *Twelve Tables*. Others find the model accurate, but rely heavily on the example of English law, where the evidence provides a better illustration of Maine's statement. Plunkett, for example, in trying to account for the 'monstrous distortion' in Bracton's *De Legibus et Consuetudinibus Angliae*, where actions constitute three-quarters of the whole treatise, resorts to the example of the *Institutes*: 'As a legal system develops more and more matter gets transferred from the law of actions to the law of obligations or of things until, finally, actions are reduced to the comparatively modest place accorded them in [Justinian's] *Institutes*.'[9]

Whatever was the true development of Roman actions, the law of actions that Gaius presents in the 2nd century AD is very much a 'residuary' division of the law in a way the law of persons and of things are not. The various components that make up the law of actions were not introduced by some superior intelligence, seeking to improve actions as a subdiscipline of the law. Instead, the law of actions comprises what is left over after centuries of picking away, of reallocation to the other divisions of the law. This means that actions is not a well-ordered subject, and instead of giving an integral whole with neat subdivisions, it frequently presents individual items of supplementary information that could not be fitted in elsewhere.

Aside from omitting discussion of formulae and actions in the law, Justinian presents a law of actions similar to Gaius'. Yet between the time of Gaius and Justinian, the idea of an action underwent a great deal of change.[10] The formulary procedure, long abolished by Justinian's time, had given a certain amount of clarity to actions: a given entry on the praetor's edict would correspond to a particular action, and as long as justice was pursued by application to the praetor for a claim from the edict, the character of individual actions would continue to be important. As the formulary procedure gave way to the new imperial procedure, however, a litigant no longer pursued a particular action, but rather presented facts which he believed would support a claim for relief under the law.

During the same period there was a broader change in the way actions were regarded. As mentioned above, in the classical law to have an action meant that a person was entitled to legal process. From there, however, it was only a short step to see an action as something that 'attached' to a person, a credit in his favour, as it was a debit against his opponent. This new conception of an action became more prevalent in the centuries between Gaius and Justinian. If an action was a 'credit' to a person, and if at the same time that action was no longer tied to a particular remedy, then an action was very much like what we regard as a 'right'.

To the Romans this new action/right seemed to have much in common with an obligation – the condition of one person owing another – and not surprisingly this results in some confusion between obligations and actions. Both the *Digest* (D.44.7) and the *Codex* (C.4.10) have titles that are headed 'On Obligations and Actions'. Also, one of the compilers of the *Institutes*, in a Greek paraphrase of the work, sought to justify obligations as an introduction to actions by remarking that 'obligations are the mother of actions' (Theophilus, *Paraphrase* 3.13). This confusion between obligations and actions continued into modern times. It was common even until the 18th century to regard the three principal divisions of the law, not as persons, things, and actions, but as persons, things (principally corporeal), and obligations and actions.[11]

Although the idea of an action had altered by Justinian's time, Justinian by and large presents the various classifications of actions as they appeared in the classical law. These classifications tell us a great deal about the law. As Gaius and Justinian present it, the law of actions asks us to take a step back from the specific actions themselves, to examine them somewhat apart from the facts under which each developed, and to describe the ways in which they differ from or resemble one another. Standing back from and analysing actions in this way results in divisions of the law quite different from the divisions presented under the law of persons and of things. For example, under the law of things we might distinguish a contract as either 'by conduct' or 'by agreement', while under the law of actions we might distinguished a contract as either 'strict law' or 'good faith'. The law of actions often ignores the substantive boundaries altogether; a restorative action, for example, might arise under either contract or delict.

The principal classifications are given below. It is one of the awkward things about the law of actions that each classification is based on a different premise. Real and personal actions are distinguished by the rights to be enforced. State-law and honorary actions are distinguished by the source of the law. Penal, restorative, and hybrid actions are distinguished by the object of the litigation. Each classification addresses a different component of the underlying law.

IV. Real and Personal Actions

(J.4.6.1 - 15, 20)

If someone were to write a new *Institutes* that expressed all of the law in the language of actions, at the top of the hierarchy would be the division between personal (*in personam*) actions and real (*in rem*) actions. All claims may be classified as one or the other, and the difference between the two types of claim is very much a matter of substantive law: a real action reflects a relationship between a person and property, and a personal action reflects a relationship between persons. The fact that a matter of substantive law is expressed as a difference in actions reflects the preference of the classical Romans for the language of remedies over the notion of rights.

A personal action arises from debt, and a real action arises from ownership. If, for example, a person receives money in payment for goods, insults another person, or damages another person's property, a debt arises between two people. If the matter comes to litigation, a claim is asserted by one person against the other. We say that the claim is personal, not because the litigation is between persons (it always is), but because the relationship being urged is one that exists between persons. Litigation over ownership, on the other hand, looks outwardly the same (person against person), but is based on something different. If a person loses possession of property that he owns and brings a claim to assert his ownership, this claim is *in rem*, because the relationship being urged is one that exists between a person and a thing.

The distinction was reflected clearly in the respective formulae. Typically the *intentio* in the formula of a real action would not mention the defendant. Instead, the issue would be framed purely in terms of the disputed ownership: Does the property belong to the plaintiff by Quiritary right (*ex iure Quiritium esse*)? Does he own the right (*ius esse*) to the fruits? In a personal action, the *intentio* typically did mention the defendant, reciting the issue with the formal language of debt: Should the defendant give (*dare oportere*) 1,000 sesterces to the plaintiff? Should the defendant do something for or give something to (*dare facere oportere*) the plaintiff on account of a prior *stipulatio* between them?

Real actions existed to enforce many kinds of ownership. Aside from the familiar example, the claim for ownership of a thing (*rei vindicatio*), there were claims for the ownership of an inheritance (*hereditatis petitio*), of a usufruct (*vindicatio usufructus*), of a right to draw water (*aquae ductus*), and many more. Real actions also existed not to enforce ownership but to deny it. An owner of land who wished to deny another's ownership of a usufruct or servitude might bring the appropriate '*actio negatoria*', a real action. In all of these real actions it is important to remember that a plaintiff does not seek the return of the thing. The judge, as always, is limited to giving a remedy in money damages, and the description 'real' refers only to the underlying relationship the plaintiff is attempting to establish. Of course as a practical matter, where the praetor inserts a special provision allowing restitution at the judge's direction, a plaintiff might have the thing restored, but this had nothing to do with the fact that the action was real.

Personal actions are described in the *Institutes* by a clever and terse piece of reasoning which Justinian borrows from Gaius (J.4.6.14; G.4.4): personal actions are only suitable for parties who deserve to get something, not for parties who own something and want it back; if you deserve to get it, *a priori* it isn't yours. The number of personal actions is of course very large, and the most familiar of them are those based on contract or delict. Among the more familiar of the remaining actions are the action against a guardian for breach of duty (*actio tutelae*, inquiring what the defendant 'ought to give or do for the plaintiff in accordance with good faith'), the action for production of a thing (*actio ad exhibendum*, inquiring whether the defendant 'ought to produce the thing'), and the action to restore a dowry (*actio rei uxoria*, inquiring whether the defendant 'ought to restore the dowry').

One of the consequences of dividing actions according to relationship is that, when a matter comes to litigation, the entire relationship, so to speak, is under review. This is clearest in the case of a real action. If an heir seeks to protect his inheritance by bringing a *hereditatis petitio*, the issue is whether the estate belongs to him under state law. If a person owns the right to channel water across certain land and his rights are interfered with, the issue is whether that person indeed owns the right to channel water. A person educated in the common law might prefer to see the issues here framed more narrowly (and in Roman terms, *in personam*), to inquire only whether someone had, for example, wrongfully interfered with the assets of the estate or the flow of the water. But to a Roman, ownership is the issue, and ownership is therefore the idea to be championed. The common lawyer might suggest further that it is a waste of resources to do anything more than try to resolve the particular issue between the parties. The Roman would answer that the waste lies on the other side, that his ownership is 'true' not only against one opponent but against the whole world, and he should not have to wait until everyone in the world sues him and loses before he can regard something as his own.

The situation is similar for personal actions, though only actions on contracts provide clear examples.[12] When a person sues on a contract, the relationship created by the contract is the subject of the action. The formula, after reciting the existence of the contract, will permit the judge to condemn the defendant 'for whatever on that account the defendant ought to give to or do for the plaintiff', or, if the matter involves a good-faith contract, 'for whatever on that account the defendant in good faith ought to give to or do for the plaintiff'. The inquiry is much broader than the particular act – for example, the failure to pay or to hand over the goods – that brought the parties to court in the first place. In other words, the basis or 'cause' underlying a contractual action *in personam* is the debt created by the contract, not the particular act or breach that brought about the dispute. This will seem unusual to a common lawyer, who is accustomed to treating the breach of a contract, not the contract itself, as the basis of a lawsuit. And the consequences of the Roman treatment are severe;

unless a party protects himself by careful pleading, his right to sue on the same contract in the future will be consumed when issue is joined, in the same way the right of his common-law counterpart to sue on the same breach is consumed.

V. State-Law and Honorary Actions
(J.4.6.3 - 13)

The distinction between state-law and honorary actions is based on the source from which the claim is derived. In a system in which a magistrate has an independent power to create new claims, claims created within that power come to be distinguished from claims that are not so created. Claims based upon the state law and unaltered by the magistrates' intervention are called 'state-law actions'. The 'state law' in this context means the *Twelve Tables* and other legislation (including interpretations of those statutes), together with the rules developed by juristic practice. Claims of the magistrates' creation are called honorary actions, and comprise actions in which the magistrate has exercised some degree of innovation, either by altering an existing state-law action (as in the Publician action) or in creating an entirely new action, not recognised as part of the state law (as in the *actio de dolo*). In Papinian's phrase, the honorary law acts 'to aid, supplement, or correct the state law, in the public interest' (D.1.1.7.1). 'Honorary' is an adjective that means 'pertaining to the office of a magistrate'; it includes the office of both the praetor and aedile.

A magistrate was called upon to innovate when an action under the state law did not speak to a particular problem or would not produce a satisfactory result, and his innovations took several forms. A common form was the 'fictitious formula', a formula that directed the judge to accept as true something that was not. The most familiar example is the Publician action: a person who had lost possession of property and was not an owner under state law might be permitted to bring what amounted to an owner's vindicatory action. It was a vindicatory action in all respects, except that it asked the judge to assume as true the falsehood that the plaintiff had satisfied the time limits of usucapion. (The Publician action was available both to bonitary owners and *bona fide* possessors, though by Justinian's time bonitary ownership did not exist, and hence he speaks only of *bona fide* possessors at J.4.6.4.) Fictitious formulae were also used to good effect in lawsuits over inheritance. The state-law rules for intestate succession operated narrowly in favour of agnates, and as a result an emancipated child, for example, would not take a share of his parent's estate as heir, as one might expect. The praetor, however, innovated aggressively in this area, and in supplementing the state-law scheme by the institution of *bonorum possessio*, allowed the emancipated child to take the estate 'as if an heir' (G.4.34; see Chapter 4, section II(c) above). Fictions such as these allowed the praetor to be innovative without disturbing the law too much:

> Roman fictions are common in two contexts, in pleadings and in legislation. Their function is the same in both, namely to extend a parcel of knowledge which is fixed and safe: we know exactly what happens when X is the case; now that Y is the case, we will proceed in exactly the same way, 'as if the case were X'. This is economical, cautious, and rigorous.[13]

A second way in which a magistrate innovated was by granting actions on the case. This type of innovation, as in the fictitious actions just described, was distinguished by a characteristic formula. But unlike a fictitious action, an action on the case might be highly creative and far-reaching.

In a state-law action, the formula was ordinarily *in ius concepta* ('conceived on the basis of the state law'). In practice this meant that the *intentio* of the formula was framed so as to recite a set of circumstances recognised in the state law. The formula in such an action would contain certain legally charged words, such as 'duty' (*oportere*), 'belong' (*rem suam esse*), 'sell' (*vendere*). In the buyer's action, for example, the *intentio* would inquire 'whereas the plaintiff bought from the defendant a thing which is the subject of this action ...'. This recites the essence of a sale under state law, that the plaintiff bought a thing. A second example is the action for non-manifest theft. Here the *intentio* would inquire 'if it appears that the theft of the thing was carried out by the defendant, for which act the defendant ought to pay a penalty as thief ...'. This language follows the offence of non-manifest theft as recited in the *Twelve Tables*.

In an action on the case, however, the *intentio* was drafted 'on the facts'. This meant that the *intentio* would simply recite certain factual allegations, and would direct the judge to condemn if he found those allegations to be true. In such a case the judge was saved the trouble of investigating whether the plaintiff's claim was made out under the rules of the state law. So, for example, the formula in an action for fraud recited certain hypothetical facts, such as 'if it appears that the plaintiff [has suffered some harm] by the fraud of the defendant ...'. If the judge found these facts to be true, he would condemn the defendant.

We can see the value in this method of innovation by considering a particular action on the case, the *actio de recepto*.[14] If a shipmaster, innkeeper, or stablekeeper undertook to keep a person's property safe, and then did not restore the property, the praetor would grant an action against him. (See Ulpian, D.4.9.1 pr, and Chapter 5, section VI above) It is surprising at first that the praetor saw fit to create this action, because there was no shortage of other actions available. An undertaking to keep something safe might constitute a *locatio conductio operis* if undertaken for pay, or *depositum* if not, and each might provide a remedy if the thing were not returned. Also, if the property were stolen or damaged, an action for theft (Ulpian, D.47.5) or damage (Paul, D.4.9.6) might be available against the keeper, even if the act were committed by an employee of the keeper. Yet the *actio de recepto* is a useful addition: where the loss was not a result of theft or damage, or the fault was not of a degree (*dolus*) to allow an *actio depositi*, a general-purpose action holding the keeper to his undertaking was desirable. It may have required a degree of accountability of the keeper that was simply not available under any other action (though whether this was true of *locatio conductio* is uncertain), and had the further advantage that it imposed an unforgiving standard of conduct on professions that were not held in high regard.

Actions on the case were particularly useful in two areas of the law. First, they were of enormous importance under the Aquilian Act. Under the statute itself the requirement of causation was fairly narrow, restricted essentially to harm that was caused directly. The introduction of actions on the case allowed an aggrieved person to have a remedy even when the harm was caused indirectly. Second, a certain class of contractual actions on the case existed, called 'actions with a special preface' (*actiones praescriptis verbis*). These were praetorian extensions of the state-law actions on obligations contracted by conduct, and through these actions the praetor could recognise the existence of transactions that did not fit within one of the traditional categories of contract. Technically (and somewhat confusingly) these extensions were regarded as state-law actions themselves, as each was modelled closely on the contract it resembled. But the formulae were drafted as actions on the case: a *demonstratio* was added at the beginning (hence 'special preface'), reciting the underlying facts of the transaction.[15]

Where a state-law action and an action on the case existed for the same underlying institution (as for the Aquilian Act), it was not always true that the action on the case was an innovation upon the state-law action. The difference, again, lay in the way the formulae were drafted, and we can take the example of the action for deposit. We know that, early on, the *Twelve Tables* allowed a penal action against a depositee under certain (unknown) circumstances. At some time during the early Republic the praetor recognised a new action on the case, the *actio depositi in factum*. As Gaius describes it (G.4.47), the action looked something like this:

> If it appears that the plaintiff deposited a thing with the defendant and through the fraud of the defendant it has not been restored to the plaintiff, condemn the defendant to pay the plaintiff whatever the thing is worth.

In creating this action the praetor probably recognised that the remedy under the *Twelve Tables* left something to be desired. Perhaps the *Twelve Tables* remedy was allowed only under narrow circumstances, or there was a need, apart from the existing penal remedy, for a new remedy allowing a depositor simply to recover the value of his property.[16] In any event, this action, with its formula drafted 'on the facts', was the result of praetorian innovation. In time the contract of deposit, with the help of juristic interpretation, came to be included among the so-called obligations contracted by conduct and gave rise to an obligation under state law. And as with the other contracts in this category (except *mutuum*), a depositor might pursue a broadly grounded 'good faith' state-law action. This action, as again Gaius describes it (G.4.47), looked something like this:

> Whereas the plaintiff deposited a thing with the defendant, whatever on that account the defendant ought to give to or do for the plaintiff in good faith, condemn the defendant.

The state-law action for deposit was therefore a later creation than the action for deposit on the case.

Because both state-law and honorary actions were administered by the same person, it is often difficult to draw a line between them. Schulz[17] mentions the example of the vindicatory action: nothing would seem to belong more to the state law than an action which states 'condemn the defendant if it appears that the disputed thing belongs to the plaintiff by Quiritary right'. And yet the additional clause added by the praetor – 'unless the thing is restored to the plaintiff according to [the judge's] direction' – utterly transforms the action. The praetor was called upon to innovate here because, without his intervention, the judge would be permitted to condemn only in money damages, and this could not be a satisfactory remedy in every case. But in allowing restitution at the judge's direction (*arbitrium*), the resulting 'discretionary action' becomes difficult to classify firmly as either state-law or honorary.

A comparison between a Roman magistrate's power to innovate and the historical equity jurisdiction of the English Chancellor is unavoidable. In many respects the distinction between the state law and the honorary law resembles that between the common law and equity in England. But there is less to this resemblance than first appears. The Roman magistrate presided over both types of action and therefore, as Buckland[18] says, 'We shall not find in the Roman law a system of rules developed gradually by a permanent tribunal whose function it was to give relief which for any reason could not be obtained in the ordinary courts.' Also, when we consider that some of the innovations introduced by the praetor out of a desire for equity would then be memorialised in his edict for future cases, the praetor resembles more a legislator than a chancellor. Finally, it is a fair argument that the greater source of equity in Roman law is not the praetor but the jurists, whose innovations would be felt principally in the second, trial phase of a lawsuit.[19]

VI. Restorative, Penal, and Hybrid Actions

(J.4.6.16 - 19; 4.12.1)

This classification is based on the object of the litigation. As the *Institutes* explains, some actions are directed to providing compensation, some to inflicting a penalty, and some to both of these things. A cynical reader may come to the conclusion that there is too much classification here. The aim of the discussion, in the end, seems to be to identify the minority of actions that are either wholly or partly penal, and to indicate the consequences of bringing such actions.

A restorative action is one which would give to the plaintiff an award that does not exceed his loss. This may seem like a roundabout way of describing something fairly simple, but in fact the definition must be put carefully. The *Institutes* gives the example of an action for the death of a slave under the Aquilian Act (J.4.6.19). Under the action the slave, in a given case, might be valued higher than his worth at death, and the difference between the two values would be regarded as penal. We might treat this as a question of valuation and view the owner of the slave as overcompensated rather than avenged, but that is not how the matter is treated in the *Institutes*. Perhaps what this discussion best illustrates is that restorative actions are defined by what they are not (neither penal nor hybrid), and that time is better spent identifying actions that are wholly or partly penal. Most actions, after all, are restorative. In general, real actions and contractual actions fall into this class, leaving delictual actions as penal or hybrid.

A penal action typically will (1) exact a sum greater than the amount of the loss, as just described, or (2) exact a multiple of the loss. The fact that they were viewed as inflicting a penalty led to certain other features. The first concerns the matter of transmissibility. The issue arises when a person who might have become a plaintiff or defendant dies before issue is joined. In such a case an heir might be permitted to bring the action the decedent would have brought, or be vulnerable to the action that would have been brought against the decedent. If the action is one that permits an heir to assert the decedent's claim, it is said to be actively transmissible. If it is one that permits an heir to have a claim asserted against him, it is said to be passively transmissible. The general rule was that restorative actions were both actively and passively transmissible, but that penal actions were only actively transmissible. The rule of course reflects the idea that only the wrongdoer himself should be punished. An exception was made for contempt: it was neither passively nor actively transmissible, in keeping with the notion that the outrage suffered by the victim of contempt belongs to him alone.

The second feature was that, if more than one person were liable under such an action, both were liable in full, so that satisfaction by one did not release the others. If for example two persons had committed a non-manifest theft, each was liable for the double penalty. The rationale is the converse of that for the previous rule: just as it makes no sense to punish someone who did not commit the act, it makes no sense to spare someone who did.

The third feature is that a penal action permitted noxal surrender. This is discussed below.

The *Institutes* glosses over what is actually at issue here, and presents the distinction between penal and restorative actions (almost) as a purely academic matter of classification. The real issue is bar. In general a person was not permitted to pursue two actions on the same matter. As soon as a dispute had passed the point at which issue was joined (the conclusion of the proceedings before the magistrate), a litigant could not raise the matter again. But this rule held true only for multiple restorative actions, or multiple penal actions. A person was permitted to bring both a restorative action and a penal action on the same matter, and hence the importance of identifying which actions were penal, wholly or in part. For example, a victim of theft could bring both an *actio furti* (penal) to punish the thief and a *rei vindicatio* (restorative) to the get the thing back, but not a *rei vindicatio* and a *condictio furtiva* (restorative).

Given the bar of multiple actions, it was particularly important to identify which actions were 'hybrid', that is, both restorative and penal.[20] A hybrid action would bar any further suit, of either type, on the same matter. One particular hybrid action both illustrates the usefulness of the classification 'hybrid' and betrays the true purpose of the restorative/penal/hybrid classification. The *Institutes* gives the action on robbery as an example of a hybrid action (J.4.6.19). It is a hybrid action because, of the fourfold penalty inflicted on the defendant, only three parts are considered to be a penalty. Centuries earlier Gaius had discussed the same subject, but unlike Justinian omitted robbery from his examples of actions 'both restorative and penal' (the term 'hybrid' not then existing). Instead, he included robbery among the purely penal actions, explaining that 'in the opinion of some' that is where it belonged (G.4.8). To say 'in the opinion of some' is tantamount to saying that in the opinion of others the action on robbery ought to be classed as both penal and restorative (it being beyond argument that anyone would class it as restorative alone).

Yet the dispute over how the action on robbery ought to be classified had nothing to do with classifications *per se*. The true issue was probably whether a robber could be treated as a 'thief' and sued by the *condictio furtiva*, a restorative action one brought against thieves. If a robber were a thief, the *condictio furtiva* would be available unless the action on robbery were deemed to be restorative in part, in which event it would bar any further restorative action. In short, what is presented as a scholarly disagreement over classifications is in fact an argument over bar, something genuinely significant to a litigant.

VII. Other Classifications
(J.4.6.28 - 30; 4.8)

There are other classifications of actions which, unlike the classifications just given, are relevant only within certain areas of the law. That they are treated under the law of actions and not in the appropriate places under persons or things may be explained by a desire to avoid repetition, or the fact that they present characteristic formulae which make them attractive to discuss as actions.

The most important of these other classifications pertains to certain personal actions. These actions were classed as to whether they were 'good-faith' or 'strict-law' (J.4.6.28 - 30). Superficially the difference was one of pleading, but the actions reflected a difference in substantive law as well. A strict-law action was characterised not by what it said but by what it did not say: the judge who presided in such an action could not consider any matters that were not a part of the pleadings, that is, the formula. If, for example, a defendant in a suit on a stipulation wished to argue that he was induced to give the stipulation by fraud or duress, or that his opponent had agreed not to pursue him, he would have to plead the matter expressly. The formula directed the judge very plainly to condemn the defendant if it appeared that the defendant ought to give a sum to the plaintiff. Accordingly, in the absence of special pleading, the judge would confine himself to examining the integrity of the stipulation. In a good-faith action, however, the judge was given far wider discretion to consider other defences, and this grant of discretion was an integral part of the formula. A formula in a good-faith action directed the judge to inquire what the defendant ought to give or do for the plaintiff in accordance with good faith. The addition of the words 'in accordance with good faith' (*ex bona fide*) distinguishes a good-faith action. The consequence of this addition is that a judge presiding, for example, in an action on sale, may absolve on his own motion a defendant who he believes was a victim of fraud when he agreed to purchase goods: such a defendant should not in good faith pay the purchase price.

Certain specific actions were set apart and classed as good-faith actions. The most important of these were actions on obligations contracted by agreement, and those contracted by conduct, with the exception of *mutuum*.

What Justinian describes as 'noxal actions' (J.4.8) might be more fully described as 'actions that allow noxal surrender as a remedy'. The remedy arises in the context of delict, and resembles a 'delictual mechanic's lien'. A person sometimes becomes liable for a delict committed by someone else, either a slave or a person within his family authority. If the delict is one that allows noxal surrender, the person liable is permitted either to pay damages or surrender the wrongdoer to the victim. By Justinian's time noxal surrender applied only to slaves, and not to children within authority.

The common explanation for the origins of noxal surrender is that liability for delict is based on revenge, and that the victim's right to avenge his loss by seizing the wrongdoer could be forgone by the payment of a ransom. Whether this is the correct explanation or not, it is consistent with certain features of the remedy. In the case of a slave, the person held liable is the person who owned the slave at the time of the action. The owner at the time of the delict is free from liability when he ceases to own the slave. Thus the aim of the action is simply to satisfy the victim, consistent with the notion of revenge. The same notion is apparent in the additional rule that a person may not have a noxal action against a slave that he owns.

The *Institutes* (borrowing from Gaius, G.4.75) justifies noxal surrender very poorly, arguing that it is unfair for a slave to inflict a loss on his owner beyond his own value (J.4.8.2). The statement is not convincing as a piece of legal analysis, and even less as a historical explanation. As Holmes points out, noxal surrender was not introduced as a vehicle for limitation of liability. His analysis is charitable: 'The Roman lawyers, not looking beyond their own system or their own time, drew on their wits for an explanation which would show that the law as they found it was reasonable'.[21]

Select Bibliography

Ankum, H., 'Gaius, Theophilus and Tribonian and the *Actiones Mixtae*', in P. Stein and A.D.E. Lewis (eds), *Studies in Justinian's Institutes in Memory of J.A.C. Thomas* (London, 1983) pp. 4-17.

Arangio-Ruiz, V., *Cour de Droit Romain: Les Actions* (Naples, 1935; repr. 1980).

Jolowicz, H.F., *Roman Foundations of Modern Law* (Oxford, 1957) pp. 75-81.

Mantovani, D., *Le formule del processo privato romano* (Como, 1992).

Stein, P., 'The Development of the Institutional System', in P. Stein and A.D.E. Lewis (eds), *Studies in Justinian's Institutes in Memory of J.A.C. Thomas* (London, 1983) pp. 151-63.

———, *Legal Institutions: The Development of Dispute Settlement* (London, 1984) pp. 127-9.

Watson, A. 'The Law of Actions and the Development of Substantive Law in the Early Roman Republic', *LQR* 89 (1973) pp. 387-92.

Wenger, L., 'The Roman Law of Civil Procedure', trans. A. Schiller, *Tul L Rev* 5 (1930) pp. 353-95.

Notes

1. See generally P. Stein, *Legal Institutions: The Development of Dispute Settlement* (London, 1984) pp. 128-9; H.F. Jolowicz, *Roman Foundations of Modern Law* (Oxford, 1957) pp. 66-81.

2. The flaw in the definition is that 'one's due' (*quod sibi debetur*) seems to exclude real actions.

3. H. Honsell, T. Mayer-Maly and W. Selb, *Römisches Recht* 4th ed. (Berlin, 1987) p. 218n2.

4. See P.G. Stein, ' "Equitable" Remedies for the Protection of Property', in P.

Birks (ed), *New Perspectives in the Roman Law of Property* (Oxford, 1989) pp. 185-94.

5. *L'année épigraphique* (1973) no. 155.

6. The exception is the quoted formula. See Nicholas, *Introduction* p. 24n2.

7. H.S. Maine, *Dissertations on Early Law and Custom* (London, 1883) p. 389.

8. A. Watson, 'The Law of Actions and the Development of Substantive Law in the Early Roman Republic', *LQR* 89 (1973) pp. 387-92. Watson generally maintains that there was a 'strict Roman separation of substantive law and procedure'. A. Watson, 'The Structure of Blackstone's Commentaries', *Yale Law Journal* 97 (1988) pp. 798, 807.

9. T.F.T. Plunkett, *Early English Legal Literature* (Cambridge, 1958) p. 51.

10. On what follows, see P. Stein, 'The Development of the Institutional System', in Stein and Lewis (eds), *Studies in Memory of Thomas* pp. 161-3, and Birks and McLeod (eds), *Institutes* pp. 17-18.

11. H.F. Jolowicz, '*Obligatio* and *Actio*', *LQR* 68 (1952) pp. 469-74.

12. On what follows, see W.W. Buckland, 'Cause of Action: English and Roman', *Seminar* 1 (1944) pp. 3-10.

13. P. Birks, 'Fictions Ancient and Modern', in N. MacCormick and P. Birks (eds), *The Legal Mind: Essays for Tony Honoré* (Oxford, 1986) p. 95.

14. On what follows, see generally R. Zimmermann, *The Law of Obligations: Roman Foundations of the Civilian Tradition* (Oxford, 1996) pp. 514-20.

15. Zimmermann, above note 14, pp. 532-5.

16. See generally R. Evans-Jones, 'The Penal Characteristics of the "*actio depositi in factum*" ', *Studia et Documenta Historiae et Iuris* 52 (1986) pp. 105-60 .

17. F. Schulz, *History of Roman Legal Science* (Oxford, 1946) p. 83.

18. W.W. Buckland, *Equity in Roman Law* (London, 1911) p. 1.

19. Id. pp. 1-8.

20. On what follows, see H. Ankum, 'Actions by Which We Claim a Thing (*res*) and a Penalty (*poena*) in Classical Roman Law', *BIDR* (3rd Ser.) 24 (1982) pp. 15-39.

21. O.W. Holmes, Jr., *The Common Law* (Boston, 1881; repr. New York, 1991) pp. 8-9.

7. Criminal Trials

O.F. Robinson[1]

I. Introduction

(J.4.18 pr, 1, 12)

Justinian devotes a part of his *Institutes* to criminal trials (*de publicis iudiciis*), but Gaius does not; for Gaius such material had no place in an outline of Roman private law.[2] Justinian indeed places it at the very end, after the title (which is also peculiar to him) on the powers and duties of judges. His purpose was explicitly to provide an introductory course on which the law student could build (see the *constitutio Imperatoriam Maiestatem* 3 - 4). It was therefore logical for him to end the title with the statement (J.4.18.12) that he had set out these notes so that the student could get a first, light contact with the material, as though flicking through the pages with his forefinger. The student would find more in the fuller volumes of the *Digest* (book 48 in particular, and also in the *Codex*, book 9) which were intended for a later stage in the law curriculum.

The title on criminal trials is also to be understood as referring back to the statement in the introductory title (J.1.1.4) that there were two sides to legal studies, public law and private law. Further, Justinian was making a connection with the main subject of his book 4 (and the sole subject of Gaius, book 4) by starting the title with a sentence about the link with actions, contrasting criminal with civil procedure (J.4.18 pr). This contrast was long outdated because of the universal jurisdiction (*cognitio*) of the urban prefect or provincial governor, but it was the history of Roman criminal law – right back to republican times – which shaped Justinian's treatment.

The term '*public* trials' (*iudicia publica*)[3] was used conventionally for criminal trials because in theory any member of the public could bring a criminal charge after the commission of an offence (J.4.18.1). There was no police force to investigate crime at Rome, although there were forces for the keeping of public order, and there was no state prosecution service. While the primitive period remains obscure and somewhat controversial, in the later Republic formal accusations made before one of the standing jury courts (*quaestiones perpetuae*) against citizens were akin to private actions in that they were brought by private individuals. These trials were called *iudicia publica* in contrast to criminal cases heard before one of the republican popular assemblies, which were called *iudicia populi*, where the charge was formally brought by a magistrate. The *iudicia publica* were public in that they were open to any adult male citizen, often, perhaps normally, without a personal stake; it was an aspect of his enjoyment of public law rights.

This was why women[4] (and also slaves,[5] to a far greater degree), who lacked rights in public law, were severely restricted in their powers of accusation. They could bring criminal charges for particularly serious offences, such as treason or manipulation of the corn supply. Otherwise women could accuse only where their own interests were involved in some way, and even then not where there was some suitable male who also had an interest. We are told that a woman could accuse a guardian of being untrustworthy; this counted as a criminal (i.e. 'public') action:

> It can be brought by anyone, even a woman, as held in a written reply of the Emperors Severus and Antoninus. But a woman can only take these proceedings when duty-bound. For example a mother can. A foster mother and a grandmother also can. And a sister too. In fact the praetor will allow any woman to make the charge if he considers that her behaviour is not an affront to the natural quietness of the female character but stems from a dutiful concern to right an injustice to the ward. (J.1.26.3).

Similar information comes from the *Codex*.[6]

II. Capital and Non-Capital Crimes
(J.4.18.2)

The stress on the distinction between capital and non-capital crimes (J.4.18.2) was also historically based. 'Capital' as applied to Roman citizens of the Republic frequently did not mean the death penalty, although it might do so on occasion. (However, it is certain that slaves were regularly put to death, as were foreigners.) The term was mostly used to describe loss of citizenship, consequent on exile either (theoretically) voluntary or enforced.[7]

(a) Banishment
(J.1.16.1 - 2, 6)

It seems to have been customary in the Republic for accused persons, at least of the upper classes, to be allowed to go into exile during their trial, before the verdict had been pronounced. After they had thus abandoned their citizenship by settling elsewhere (for dual nationality was not a received concept in the ancient world, G.1.131) they were interdicted from fire and water, hearth and home. This meant that they were outside the law and, if they returned, had no civic rights and indeed could summarily be put to death. Loss of citizenship was not itself the penalty, but the inevitable consequence of the exile.

The political changes of the Principate meant, first, that it became repugnant to imperial authority that a man should escape the formal penalty for his misdeeds by an act of his own volition and, second, that because direct control of the provinces steadily replaced self-government, except at local level, it was no longer really possible to move away from Roman territory. (The process was complete after the extension of the citizenship to virtually all the free inhabitants of the Empire by the emperor Caracalla's enactment of ca. AD 212.) We therefore find two forms of banishment being formalised. 'Deportation' was capital. It meant sending someone to a particular and fairly unpleasant place, usually an island or the Great Oasis (e.g., J.1.12.1); it necessarily involved loss of citizenship and was of its nature perpetual, only capable of being brought to an end by an imperial pardon. 'Relegation', in contrast, was the removal of someone from Rome and, normally, his home province, although he might be directed to stay in a particular place, as the poet Ovid was sent to Tomis on the shores of the Black Sea; such a man retained authority over his children (J.1.12.2). It was not capital since, although it might be accompanied by some confiscation of property, it was not held to be a commutation of the death penalty, as was deportation. These terms had been precise in the Principate, but Justinian used them with some freedom when describing status-loss (*capitis deminutio*) (J.1.16.1 - 2, 6; see Chapter 2, section IV(b) above).

However, although certainty is impossible, it seems likely that even in the Republic persons of the lower orders who were caught in the act of committing some serious crime might not always have been given the opportunity to assert their citizenship and go into exile but instead would have been put to death.[8] Exceptions were also made when public safety in the wider sense was involved, as with the executions after the disclosure of the Bacchanalian conspiracy in 186 BC,[9] or the putting to death of Lentulus and other Catilinarian conspirators.[10] In the course of the Principate penalties became more strictly administered, and actual death began to be inflicted regularly, even on citizens. Ingenious forms of the aggravated death penalty also became common.[11]

(b) Upper and lower classes

In the Republic the important distinction had been between citizens and foreigners – slaves counting below all free men. In the 2nd century AD there emerged a formalised distinction between the upper and lower classes of society, the *honestiores* and the *humiliores*.[12] The former included the members of the senatorial and equestrian orders, town councillors (decurions) (each of these classes was defined by a compound of birth, office, and wealth), practitioners of certain professions, and also honourably discharged veterans, all with their families. Most other people fell into the class of *humiliores*. The distinction was of importance in determining personal liability for the municipal burdens which became so important in the later Empire, and also in determining the weight of a man's evidence, but its effects were mainly apparent in the criminal law. First, a member of the *humiliores* came to be liable to torture, not only when a suspect but also merely as a witness. Second, after conviction the penalties were normally different for the two classes, being harsher for *humiliores*; indeed in this respect they became largely equated with foreigners of the republican period. However, the distinction was flexible; certain groups, such as physicians, might be poised between two classes, and slaves were often but not always treated as a third class. Where *humiliores* were put to death elaborately, by burning or crucifixion (before the Empire became Christian) or being sent to the beasts (on conviction, for example, for fireset-ting in Rome or murder), *honestiores* would normally suffer simple death by decapitation. For forgery, where *humiliores* would be condemned to simple death or to penal slavery in the mines, *honestiores* would suffer deportation.

(c) Penal slavery
(J.1.12.3; 1.16.1; 3.12.1)

'Penal slaves are those sentenced to the mines or thrown in with wild beasts' (J.1.12.3). Those sentenced to work in the mines (*in metallum*), unlike those condemned to forced labour (*opus publicum*), were automatically penal slaves, as were those condemned to fight in the hunting-games.[13] Condemnation to the various forms of the death sentence also left the criminal a penal slave while awaiting execution. Normally a capital sentence involving penal slavery was accompanied by confiscation of the convict's entire patrimony, although this was sometimes partially remitted in the interest of his dependants.[14]

Loss of liberty was on its own an ancient punishment, found in the *Twelve Tables* of the 5th century BC, but by the end of the Republic it had virtually ceased to exist as a specific penalty. Slavery as a quasi-penalty was imposed on freedmen who were reduced to their former status for gross ingratitude to their patron, and on a man who allowed himself to be sold as a slave in order to share the price (J.1.16.1), but such persons were really being confirmed in their servile status. Enslavement under the Claudian Resolution of the Senate of AD 54 might also be deemed to fall into this category. This resolution was applicable to a free woman who persisted in cohabiting with a slave against the wishes of his owner (G.1.91), and probably originated in cases of cohabitation with imperial slaves, whose status made them attractive, where the emperors wished to make sure that any offspring stayed bound to their service.[15] Justinian viewed it as shocking, as well as declaring it obsolete (J.3.12.1).

(d) Lesser penalties
(J.1.12.2; 1.16.5; 1.26.8)

Penalties other than death, deportation, the mines, and the arena were not capital. Someone relegated kept his or her status as a citizen, and the punishment, although it might be indefinite, was commonly temporary. Relegation we have contrasted with deportation, but both were penalties which seem more commonly inflicted upon the upper classes, whereas their equivalents for the lower orders were forced labour and the mines. Forced labour,[16] perhaps in the public baths or public bakeries, was imposed on *humiliores* where relegation was the penalty for *honestiores*, for example, after conviction for cattle-stealing. Confiscation of property, at least partial, often accompanied these sentences, but there was no formal status-loss. The most common lesser penalty was a flogging for *humiliores*, a fine for *honestiores*. In the later Empire mutilation was sometimes found beside flogging as a form of corporal punishment;[17] Justinian in general forbade its use,[18] but it reappeared in the Byzantine Empire.

Infamy – involving, for example, inability to act as a formal witness or to appear in court on another's behalf – was sometimes a criminal sentence and sometimes a consequence of being condemned in a private suit in an action based on good faith; the latter ground merged with the former in a process of criminalisation during the course of the 3rd century. Some penalties of this kind were by their nature restricted, such as removal from the senate (J.1.16.5) or from the class of town councillors (*ordo decurionum*), or the withdrawal of an advocate's licence to practice. Women could not suffer this kind of penalty because they lacked the ability in the first place. *Humiliores* were not perceived as having a public standing, or a reputation to lose.

Imprisonment was not itself a general penalty in the Empire; its use was in theory merely custodial until execution or despatch to the mines, or to the place of exile or of forced labour. However, where a fine was imposed and adequate security not given for its payment, the condemned person might be held in prison. We do hear of cases where a death sentence was effectively turned into a sentence of imprisonment by the indefinite postponement of the execution,[19] but Ulpian says, definitively at least within the context of the standing jury courts, that prison was for custody not punishment.[20] Chains (*vincula*) were also custodial or coercive rather than penal. Nevertheless, the restraint imposed on someone condemned to forced labour seems not so different from this century's prison with hard labour.

The death of an accused normally brought a trial to an end (J.1.26.8), although his heirs would be liable for any (unjustified) enrichment of the estate. The one clear exception was in trials for treason, where a dead man could be condemned, and his family might thereby be tainted.

III. Specific Offences
(J.4.18.3 - 11)

The list that Justinian gives (J.4.18.3 - 11) of specific offences is also historically based. It reflects the crimes that were legislated for by the introduction of standing jury courts in the late Republic and under the emperor Augustus (27 BC – AD 14). These crimes (treason, sedition, embezzlement of public funds, improper application of public funds, extortion, electoral corruption, interference with the corn supply, adultery, murder, parricide, violence, kidnapping, and forgery) were all the creation of a regulatory statute. Statute and interpretation elaborated substantive offences and extended the sphere of particular courts.

Treason[21] (*Lex Iulia maiestatis*, J.4.18.3) ranged from direct betrayal of Romans to an enemy, to behaviour insulting (perhaps only mildly) to an emperor or a member of his family; its various forms can be defined as high treason (*perduellio*) and lèse-majesté (*crimen laesae maiestatis*[22]). It thus covered armed rebellion, failure to relinquish one's province after the arrival of a successor, plots to kill magistrates or other persons in authority, desertion from the army, making false entries in public records, melting down consecrated statues of the emperor, writing lampoons about the emperor or members of his immediate family, and even coining when it could be held to have abused the imperial image.

Adultery[23] (*Lex Iulia de adulteriis coercendis*, J.4.18.4) was, strictly speaking, fornication by or with a married woman, but the crime was extended to cover most sexual irregularities, such as rape (though this could also fall under force or violence), abduction,[24] the seduction of respectable women who were widowed or virgin, male homosexual practices (except with slaves), and pimping. Bigamy was not an offence until the institution of marriage became a formal legal concept under the Christian Empire (compare Chapter 2, section III above). Adultery was repressed more severely in the Christian Empire than in the classical period, when relegation seems to have been the normal penalty.[25] Incest does not seem to have fallen directly under the Julian Act on adultery, although it was seen as a related offence.[26] In the *Institutes* its treatment is more important in the context of a bar to marriage (J.1.10.1 - 10, 12; G.1.59 - 64); Gaius did not even refer to the possible criminal implications.

Murder (J.4.18.5), which as a statutory offence originated in brigandage (*Lex Cornelia de sicariis*), came to cover abortion and even castration, as well as deliberate setting of fires, and killing by poison or magic. It seems that this area of the law was relatively undeveloped,[27] partly, one suspects, because the death of a slave would often simply be the subject of an Aquilian action. On the other hand, the overt murder of a slave-owner, a term which would presumably cover the majority of free persons, was governed by the Silanian Resolution of the Senate.[28]

Parricide[29] (J.4.18.6) is very rarely heard of in the historical sources, reasonably enough. The aggravated penalty (being sewn into a sack with a dog, cock, snake, and monkey, and thrown into the river or sea) seems only to have been imposed for the murder of ascendants, mother as well as father, grandparents as well as parents. While the term was used for the killing of other close kin, as far as first cousins and including step-parents, children, spouses, and in-laws, the penalty then was the same as for ordinary murder, that is, the simple death penalty or relegation, depending on status. The introduction of various animals into the sack was probably the work of the emperor Constantine (AD 312-37) or his sons. Justinian's interest in parricide seems quite disproportionate to its importance; he may simply have been fascinated by the picturesque penalty, or have thought that students would be.

Forgery (*Lex Cornelia de falsis*, J.4.18.7) was a two-headed crime. We hear of a statute (*Lex Cornelia nummaria*) which dealt with counterfeiting money, producing false coins, or shaving or otherwise reducing the value of good ones.[30] There was also a *Lex Cornelia testamentaria* which was concerned with forging wills, whether by addition or subtraction, and also with tampering with other documents.[31] These two aspects were treated together by the jurists, and presumably by the courts, of the classical period, but were separated in Justinian's codification.

Violence or *force*[32] (*Lex Iulia de vi*, J.4.18.8) could cover anything from sedition or riot to assault or forcible dispossession. Rape could also fall under this statute, and not only of virgins or widows. Its sphere overlapped the crime of contempt (*iniuria*), which comprised both assault and defamation; this had been regulated by a Cornelian Act and was treated there as a crime as well as a delict. The existence of the delict seems to have led to a relative lack of interest in the crime.

Embezzlement of public funds[33] (*lex Iulia peculatus*, J.4.18.9) was an offence going back at least to Sulla. It involved unlawful acquisition, in effect theft, from the state; it was extended to cover tampering with public records. *Sacrilege* and the *embezzlement of monies remaining* had been tried before the same jury court as *peculatus* under Augustus; Justinian held them, in contrast, non-capital. The embezzlement of monies remaining (*lex Iulia de residuis* according to J.4.18.11, but perhaps not a separate act) meant the improper application of public funds of which a man had come into lawful possession. Sacrilege in classical law had been simply an offence of dishonesty directed against money or property dedicated to the gods; under Justinian it was an offence against religion.[34]

Kidnapping[35] (*Lex Fabia de plagiariis*, J.4.18.10) applied not only to free people but also to slaves, when the motive was lust or revenge or some other non-lucrative purpose; the essence of the crime was usurpation of a right, whether of ownership or of paternal power. In the later Empire it acquired a new slant when agricultural tenants came to be tied to their farms, an early form of serfdom; it was linked with the abuses of the powerful who operated a system of illegal forced labour. The Fabian Act on kidnapping was not itself tied to the creation of a standing jury court but went back to the late 3rd or early 2nd century BC.

Electoral corruption (*Lex Iulia ambitus*, J.4.18.11) had been a much-used criminal charge in the late Republic; it was a flourishing aspect of roughly-played party politics.[36] Its importance diminished steadily during the Empire as autocracy replaced oligarchy, and for Justinian it was only relevant to municipal elections, or promotions within the civil service. *Extortion*[37] (*Lex Iulia* [*rerum*] *repetundarum*, ibid.) originally described what was seized or exacted by force or blackmail from citizens residing in the provinces, and then from their native inhabitants, by the Roman governors. It was extended to cover everything improperly extorted by anyone in official power, or by any member of his family. *Interference with the corn supply*[38] (*Lex Iulia de annona*, ibid.) included both physical and administrative misbehaviour, for instance, using false measures or manipulating grain prices.

Justinian confines his list almost entirely to the statutory crimes created with, and for, a standing jury court. The field of delict covered much minor crime, particularly theft, even in the later Empire. It is clear that there was a role for a common law, that is, for interpretation based on the *Twelve Tables*, in criminal matters;[39] murder was certainly an offence before passage of the Cornelian Act. After the establishment of the Principate, interpretation – particularly by resolution of the senate – extended the scope of most statutory crimes. There remained, however, a category of *crimina extraordinaria*, crimes outside the list based on statute. These might include such offences against morality and good order as gambling or astrology, or removing bodies or bones from tombs;[40] they also included offences against religion, whether committed by druids or Christians.[41] There were also procedural offences, such as calumny,[42] and other doubtless rarer and more curious cases, such as the pedlars who carried around snakes and produced them, apparently to frighten bystanders into paying up.[43] These and other minor offences[44] were dealt with administratively rather than by due process.

IV. Criminal Courts[45]

Justinian, hardly surprisingly, did not explain the historical basis of his treatment. Before the time of Sulla, consul in 88 and dictator 82 – 79 BC, it may well be that the majority of criminal trials took place before one of the assemblies of the Roman people, of which there were several (see Chapter 1, section III(a) above).[46] Sulla systematised the use of the standing jury courts for most major offences. Any Cornelian Act in the context of criminal law, those for example on murder or forgery, is named after him, since he will have brought about its enactment by an assembly of the people. His work was refined by Julius Caesar (d. 44 BC) and Augustus, who gave their family name to various Julian Acts (for example, on treason, adultery, public or private violence, electoral corruption, the theft or conversion of public monies, extortion, and offences connected with the subsidised grain supply), although they had not necessarily brought about the original establishment of all these courts. The Pompeian Act on parricide dates from either 70 or 55 BC, although its occasion is unknown.

Parricide had roots in primitive Roman law, but seems always to have been dealt with as a special case; it was never one of the 'ordinary' crimes, even although charges were heard before the murder court. The standing jury courts set a pattern of procedure in the last century BC, but they appear to have diminished in significance almost from the beginning of the Empire. They had probably disappeared before the end of the 1st century, or at the latest by the early 3rd century,[47] but certainty is not possible.

The senate[48] became a court of some importance in the course of the earlier Principate. The legal basis for this evolution is not clear, but in some ways it was not startlingly innovative. Senators had provided a substantial proportion of the jurors for the standing courts; these juries were drawn exclusively from the upper ranks of society (and their precise composition had been a cause of political strife for some 50 years in the period after the Gracchan law of 122 BC).[49] The senate could also be viewed as the advisory body (*consilium*) of the emperor, its members acting as assessors with him, or being suitable deputies for him to appoint. Moreover, the offences of electoral corruption, extortion, embezzlement, and improper application of public funds, and many forms of treason, were particularly likely to be committed by men of senatorial rank because it was they who held the magistracies, the official positions of which the unscrupulous could take advantage.

The jurisdiction of the emperor himself was naturally of ever-increasing importance. The reorganisation of the administration of the City, under Augustus and his successors, also gave jurisdiction to new officials, who were in effect imperial deputies, such as the urban prefect, the prefect of the night watch, the prefect of the corn supply, and other commissioners. The jurisdiction of provincial governors rapidly came to be seen as something delegated by the emperor. Towards the end of the Principate the praetorian prefecture came to act as the normal tribunal representing the emperor. In the later Empire, all justice sprang from the emperor, but he had almost entirely ceased to exercise it personally.

V. Procedural Changes

This centralisation of justice had other effects. One of these was to bring slaves within the normal sphere of the criminal law, not merely making them liable for crimes they committed but also removing them in more serious cases (but presumably not in matters within the household) from their owners' domestic jurisdiction.[50] Conversely, the deliberate killing of a slave came to be comprised in the crime of murder, even, in theory, when carried out (without lawful reason) by his owner (J.1.8.2; G.1.53).

Similarly, and probably more effectively, the domestic jurisdiction of the *paterfamilias* (the father of the family, see Chapter 2, section II(a) - (c) above) was limited. The disappearance of marital subordination and, effectively, of bondage, had left only paternal power as authoritative in the developed Principate; it still included the right to put someone in one's power to death, for good cause, as late as the reign of the emperor Hadrian (AD 117-38).[51] Even the courtesy of leaving women to be dealt with by their families rather than brought before the public courts seems largely to have ceased in the 2nd century; for example, our evidence is plentiful for the public execution of female as well as male Christian martyrs in the 3rd century.[52] For political reasons it was inevitable that the state would reserve to itself the power of life and death. The state under the emperor had replaced the 'federal' grouping of agnatic families, each under its head. In the later Empire there was legislation[53] specifically restricting powers to chastise not only slaves but also near relatives, a category which presumably included freedmen as well as children and grandchildren.

Another major change between the classical law and that of Justinian was in the actual procedure of the criminal courts. In the trials before the assemblies under the Republic, and then before the standing jury courts, oratory was more important than making technical legal points. We do not know how many citizens customarily attended the assemblies which met to sit as courts; such trials had effectively ceased when Cicero (d. 43 BC) was a very young man. He, however, does tell us much, both explicitly and by implication, about the procedure in the standing courts, with their juries of some 30 to 90.[54] Claims concerning someone's character seem to have formed the most convincing testimony; we find this confirmed from Aulus Gellius' account of his experience of the civil courts.[55] This seems to have remained true in the classical period of law, for Justinian cites (J.1.26.5) Julian's opinion that a guardian could be removed as untrustworthy even before he had begun to act, at which time there could hardly be grounds proving him unsuitable for the particular duty. Nevertheless, under the impulse of imperial organisation, and indeed through legislation, rules on evidence became more formal, although the judge had discretion to control what came before the court.[56] In the later Empire trials were held in closed (or closable) halls instead of open public places such as the Forum. All proceedings were regularly recorded in writing, as was the sentence promulgated.

VI. Summary

It is very clear that, for both his list of offences and the recitation of their penalties, Justinian adhered to the patterns set in the last century BC. This is partly because that had been the most creative period in the giving of structure to Roman criminal law, and in the shaping of its courts and their jurisdiction. Rather strangely, we do not know of much juristic interest in the substantive criminal law until the Antonines in the late 2nd century AD.[57] Nevertheless, the earlier period provided the model which, although modified fairly thoroughly in detail by later generations, seems to have continued to shape their thinking.

The courts of Justinian's day had powers to deal with anything brought before them. The judges had almost unfettered discretion but, paradoxically, at the same time they were subject to close controls emanating from the emperor and his ministry of justice, or such was the theory. Yet the importance of the old ways remained great. This may have been due in part to the nature of legal education, which was based on the work of the jurists of the classical period who, in turn, looked back to their predecessors in the late Republic (the '*veteres*', though Justinian uses this term for the classical jurists themselves). It may also have been the result of a more general legal conservatism. Further, Justinian was trying to restore the glories of the legal past, and so he naturally looked there for a model. His title does not take the most logical approach to the subject of criminal law, but perhaps he only wanted to whet the appetite of students, and this may explain his stress on the bizarre penalty of the sack. The title remains a somewhat curious appendix to his *Institutes*.

Select Bibliography

Bauman, R.A., *Crime and Punishment in Ancient Rome* (London, 1997).

Beinart, B., 'Crime and Punishment in an Historical Setting', *AJ* (1975) pp. 5-39.

Cloud, J.D., 'The Constitution and Public Criminal Law', in J.A. Crook et al. (eds), *Cambridge Ancient History* IX 2nd ed. (1994) pp. 491-530.

Jones, A.H.M., *Studies in Roman Government and Law* (Oxford, 1960).

———, *The Criminal Courts of the Roman Republic and Principate* ed. J. Crook (Oxford, 1972).

Lintott, A.W., '*Provocatio*. From the Struggle of the Orders to the Principate', in *ANRW* I.2 pp. 226-67.

MacMullen, R., 'Judicial Savagery in the Roman Empire', *Chiron* 16 (1986) pp. 147-66.

Robinson, O.F., *Criminal Law in Ancient Rome* (London, 1995).

Strachan-Davidson, J.L., *Problems of the Roman Criminal Law* (Oxford, 1912).

Thomas, J.A.C., '*Sutor ultra crepidam*', *JR* (1962) pp. 127-48.

———, 'Prescription of Crimes in Roman Law', *RIDA* (3rd ser.) 9 (1962) pp. 417-30.

———, 'The Development of Roman Criminal Law', *LQR* 79 (1963) pp. 224-37.

Notes

1. An earlier version of this chapter, with more references to literature not in English, was published as 'Some Thoughts on Justinian's Summary of Roman Criminal Law', *BIDR* (3rd ser.) 33-34 (1991-92) pp. 89-104. I must thank Professor Mario Talamanca for permission to use the same material.

2. On the differences between the aims of Gaius and Justinian in their respective *Institutes*, see O.F. Robinson, 'Public Law and Justinians's Institutes', in Stein and Lewis (eds), *Studies in Memory of Thomas* pp. 125-33.

3. The term 'iudicium' is complicated by having a range of overlapping meanings: 'legal proceedings', 'case', 'trial', 'judgment', 'court', even 'jurisdiction'.

4. O.F. Robinson, 'Women and the Criminal Law', *Annali della facoltà di giurisprudenza di Perugia* (new ser.) 8 (1985) (= *Raccolta di scritti in memoria di*

R. Moschella) pp. 527-60; cf. S. Dixon, '*Infirmitas sexus*: Womanly Weakness in Roman Law', *TvR* 52 (1984) pp. 343-71.

5. O.F. Robinson, 'Slaves and the Criminal Law', *ZSS* (rom. Abt.) 98 (1981) pp. 213-54.

6. E.g. C.9.1.5 (AD 222); 9.1.12 (AD 293).

7. G. Crifo, *L'esclusione dalla città* (Perugia, 1985).

8. In the provinces there was little restraint on the powers of governors; e.g. Verres and the unfortunate Gavius, see Cic. II *in Verr*. 5.160-4.

9. For the story, see Livy 39, 8-19; also *Senatusconsultum de Bacchanalibus* (in S. Riccobono (ed), *Fontes Iuris Romani Antejustiniani* I (Florence, 1941) no. 30, pp. 240f), translated in A. Johnson, P. Coleman-Norton and F. Bourne (eds), *Ancient Roman Statutes* (Austin, TX, 1961) no. 28. Cf. R.A. Bauman, 'The Supression of the Bacchanals: Five Questions', *Historia* 39 (1990) pp. 334-48.

10. Sall. *Cat*. 55.

11. O.F. Robinson, 'Crime and Punishment and Human Rights in Ancient Rome', *RIDA* (3rd ser.) 44 (1997), supplement (forthcoming); and see K.M. Coleman, 'Fatal Charades; Roman Executions Staged as Mythological Enactments', *JRS* 80 (1990) pp. 44-73.

12. P. Garnsey, *Social Status and Legal Privilege* (Oxford, 1970); R. Rilinger, *Humiliores-Honestiores* (Munich, 1988); F.J. Navarro, *La formación de dos grupos antagónicos en Roma:* honestiores *y* humiliores (Pamplona, 1994).

13. D.48.19.8.4 - 6, 10 - 11 (Ulpian, *Duties of Proconsul*, book 9).

14. E.g. C.9.49.9 (AD 396); cf. the case of Gaius Gracchus' widow, in D. Daube, 'Licinnia's Dowry', *Studi in onore di B. Biondi* I (1965) pp. 199-212.

15. P.R.C. Weaver, *Familia Caesaris* (Cambridge, 1972) pp. 162-9. See Chapter 2, section I(a).

16. F. Millar, 'Condemnation to Hard Labour in the Roman Empire from the Julio-Claudians to Constantine', *Papers of the British School at Rome* 52 (1984) pp. 124-47.

17. E.g. *Theodosian Code* 1.16.7 (AD 331); cf. C.P. Jones, 'Stigma: Tattooing and Branding in Antiquity', *JRS* 77 (1987) pp. 139-55.

18. Nov.134.13 (AD 556), but castration was the fitting punishment for the crime of castrating another, Nov.142 (AD 558).

19. Tac. *Ann*. 6.23.

20. D.48.19.8.9 (Ulpian, *Duties of Proconsul*, book 9); but see A. Lovato, *Il carcere nel diritto penale romano* (Bari, 1994).

21. As a starting point, R.A. Bauman, *The crimen maiestatis in the Roman Republic and Augustan Principate* (Johannesburg, 1967); id., *Impietas in principem* (Munich, 1974).

22. 'Crimen' is another term that is rather confusing because of its range of meanings: 'accusation', 'indictment', 'offence', 'guilt'.

23. L.F. Raditsa, 'Augustus' Legislation concerning Marriage, Procreation, Love Affairs and Adultery', in *ANRW* II.13 pp. 278-339.

24. J. Evans-Grubb, 'Abduction Marriage in Antiquity', *JRS* 79 (1989) pp. 59-83.

25. *Paul's Sentences* 2.26.14.

26. D.48.18.4 (Ulpian, *Disputations*, book 3); 48.18.5 (Marcian, *Institutes*, book 2); cf. *Paul's Sentences* 2.26.15.

27. But see D. Nörr, '*Causam mortis praebere*' in MacCormick and Birks (eds), *The Legal Mind* pp. 203-17.

28. See Robinson, op. cit. above note 5; W.W. Buckland, *The Roman Law of Slavery* (Cambridge, 1908) pp. 94-7. See also J.G. Wolf, *Das Senatusconsultum Silanianum und die Senatsrede des C. Cassius Longinus* (Heidelberg, 1988).

29. J.D. Cloud, '*Parricidium*', *ZSS* (rom. Abt.) 88 (1971) pp. 1-66; D. Briquel, 'Sur le mode d'exécution en cas de parricide et en cas de *perduellio*', *Mélanges d'archéologie et d'histoire de l'École Française de Rome* 92 (1980) pp. 87-107.

30. P. Grierson, 'The Roman Law of Counterfeiting', in R.A.G. Carson and C.H.V. Sutherland (eds), *Essays in Roman Coinage presented to H. Mattingly* (London, 1956) pp. 240-61.

31. O.F. Robinson, 'An aspect of *falsum*', *TvR* 60 (1992) pp. 29-38; F. Marino, 'Il falso testamentario nel diritto romano', *ZSS* (rom. Abt.) 105 (1988) pp. 634-63.

32. A.W. Lintott, *Violence in Republican Rome* (Oxford, 1968); see also J.D. Cloud, '*Lex Iulia de vi*', *Athenaeum* 66 (1988) pp. 579-95; 67 (1989) pp. 427-65.

33. F. Gnoli, *Ricerche sul crimen peculatus* (Milan, 1979).

34. O.F. Robinson, 'Blasphemy and Sacrilege', *IJ* (new ser.) 8 (1973) pp. 356-71.

35. R. Lambertini, *Plagium* (Milan, 1980).

36. E.S. Gruen, *Roman Politics and the Criminal Courts 149-78 BC* (Cambridge, MA, 1968); A.W. Lintott, 'Electoral Bribery in the Roman Republic', *JRS* 80 (1990) pp. 1-16.

37. J.S. Richardson, 'The Purpose of the *lex Calpurnia de repetundis*', *JRS* 77 (1987) pp. 1-12. Most scholarly argument centres on the republican courts; see e.g. A.W. Lintott, 'The *leges de repetundis* and Associated Measures in the Republic', *ZSS* (rom. Abt.) 98 (1981) pp. 162-212.

38. The literature on the corn supply is enormous; among the more recent: H. Pavis d'Escurac, *La Préfecture de l'annone* (Rome, 1976); G. Rickman, *The Corn Supply of Ancient Rome* (Oxford, 1980); P. Garnsey, *Famine and Food Supply in the Graeco-Roman World* (Cambridge, 1988); A.J.B. Sirks, *Food for Rome* (Amsterdam, 1991).

39. B. Santalucia, *Diritto e processo penale nell' antica Roma* (Milan, 1989) ch. 4.

40. D.11.5; *Collatio legum Mosaicarum et Romanarum* 15.2; D.47.12.11 (Paul, *Sentences*, book 5).

41. H. Last, 'Rome and the Druids', *JRS* 39 (1949) pp. 1-5; O.F. Robinson, 'The Repression of Christians in the Pre-Decian Period: A Legal Problem Still', *IJ* (new ser.) 25-27 (1990-92) pp. 269-92.

42. See J.G. Camiñas, 'Le *crimen calumniae* dans la *lex Remmia de calumniatoribus*', *RIDA* (3rd ser.) 37 (1990) pp. 117-33; O.F. Robinson, *Criminal Law in Ancient Rome* (London, 1995) pp. 99-103.

43. C.9.46; D.47.11.11 (Paul, *Sentences*, book 1).

44. E.g. hooligans might be barred from attending the Games, D.48.19.28.3 (Callistratus, *Judicial Examinations*, book 6); cf. O.F. Robinson, *Ancient Rome: City Planning and Administration* (London, 1992; pbk. 1994) pp. 196f.

45. A.H.M. Jones, *The Criminal Courts of the Republic and Principate* (Oxford, 1972). One of Jones' concerns was to reconsider the challenge to the views of T. Mommsen, *Römisches Strafrecht* (Leipzig, 1899) – still the fundamental book on Roman criminal law in general – made by W. Kunkel, *Untersuchungen zur Entwicklung des römischen Kriminalverfahrens in vorsullanischer Zeit* (Munich, 1962).

46. On the possibilities of criminal procedure using a private criminal action, see W. Kunkel, 'Ein direketes Zeugnis für den privaten Mordprozess', *ZSS* (rom. Abt.) 84 (1967) pp. 382-5.

47. P. Garnsey, 'Adultery Trials and the Survival of the *quaestiones* in the Severan Age', *JRS* 57 (1967) pp. 56-60.

48. R.J.A. Talbert, *The Senate of Imperial Rome* (Princeton, 1984) ch. 16 ('The Senatorial Court'); F. Arcaria, *Senatus censuit* (Milan, 1992); O.F. Robinson, 'The

Role of the Senate in Roman Criminal Law during the Principate', *Journal of Legal History* 17 (1996) pp. 130-43.

49. M.T. Griffin, 'The *leges iudiciariae* of the Pre-Sullan Era', *CQ* 23 (1973) pp. 108-26.

50. For the powers of a republican owner, see Cicero's denunciation of Sassia (*pro Cluentio* 63.176 - 66.187); see also Robinson, op. cit. above note 5.

51. D.48.9.5 (Marcian, *Institutes*, book 14).

52. H. Musurillo, *The Acts of the Christian Martyrs* (Oxford, 1972).

53. *Theodosian Code* 9.12.1 (AD 326/329); 9.13.1 (AD 365/373).

54. A.H.J. Greenidge, *The Legal Procedure of Cicero's Time* (London, 1901) gives a very full account of the Ciceronian evidence.

55. Gell. *NA* 14.2.

56. *Theodosian Code* 11.39; C.4.19 - 21.

57. L. Fanizza, *Giuristi, Crimini, Leggi nell'eta degli Antonini* (Naples, 1982). It seems, however, highly probable that juristic advice lay behind the considerable senatorial development of the various statutes in the earlier Principate. See Robinson, op. cit. note 48.

General Bibliography

General Textbooks

Buckland, W.W., *The Main Institutions of Roman Private Law* (Cambridge, 1931).
———, *A Manual of Roman Private Law* 2nd ed. (Cambridge, 1939).
———, *A Textbook of Roman Law from Augustus to Justinian* 3rd ed. rev. P. Stein (Cambridge, 1963).
Borkowski, A., *Textbook on Roman Law* 2nd ed. (London, 1997).
Kaser, M., *Roman Private Law* 2nd ed. trans. R. Dannenbring (London, 1968).
Leage, R.W., *Roman Private Law* 3rd ed. rev. A.M. Prichard (London, 1961).
Lee, R.W., *The Elements of Roman Law* 4th ed. (London, 1956).
Muirhead, J.S., *An Outline of Roman Law* (London, 1947).
Nicholas, B., *An Introduction to Roman Law* (Oxford, 1962).
Schiller, A.A., *Roman Law* (New York, 1978).
Schulz, F., *Classical Roman Law* (Oxford 1951; repr. 1961).
Thomas, J.A.C., *Textbook of Roman Law* (Amsterdam, 1976).
Thomas, P.J., *Introduction to Roman Law* (Boston, MA, 1986).
Van Warmelo, P., *An Introduction to the Principles of Roman Civil Law* (Cape Town, 1976).

Institutional Works and Commentaries

Birks, P. and McLeod, G. (eds), *Justinian's Institutes* (London, 1987; repr. 1994).
Gordon, W.M. and Robinson, O.F. (eds), *The Institutes of Gaius* (London, 1988).
Moyle, J.B. (ed), *Imperatoris Iustiniani Institutionum* 5th ed. (Oxford, 1912).
Thomas, J.A.C. (ed), *The Institutes of Justinian* (Amsterdam, 1975).
de Zulueta, F. (ed), *The Institutes of Gaius* (Oxford, 1946, 1953) 2 vols.

History, Sources, and General and Comparative Roman Law Works

Birks, P. (ed), *New Perspectives in the Roman Law of Property* (Oxford, 1989).
Buckland, W.W., *Equity in Roman Law* (London, 1911).
Buckland, W.W. and McNair, A.D., *Roman Law and Common Law: A Comparison in Outline* 2nd ed. rev. F.H. Lawson (Cambridge, 1965).
Cracknell, D.G., *Roman Law: Origins and Influence* (London, 1990).
Crawford, M.H. (ed), *Roman Statutes* (*Bulletin of the Institute of Classical Studies*, suppl. 64) (London, 1996), 2 vols.
Crook, J., *Law and Life of Rome* (Ithaca, NY, 1967; repr. 1991).
Daube, D., *Roman Law: Linguistic, Social and Philosophical Aspects* (Edinburgh, 1969).
Evans-Jones, R. (ed), *The Civil Law Tradition in Scotland* (Edinburgh, 1995).
Frier, B.W., *The Rise of the Roman Jurists* (Princeton, NJ, 1986).
Honoré, A.M., *Emperors and Lawyers* 2nd ed. (London, 1994).
Jolowicz, H.F., *Roman Foundations of Modern Law* (Oxford, 1957).
Jolowicz, H.F. and Nicholas, B., *Historical Introduction to the Study of Roman Law* 3rd ed. (Cambridge, 1972).

General Bibliography

Kunkel, W., *An Introduction to Roman Legal and Constitutional History* 2nd ed. trans. J.M. Kelly (Oxford, 1973).
Lambiris, M., *The Historical Context of Roman Law* (North Ryde, NSW, 1997).
Lawson, F.H. (ed), *The Roman Law Reader* (Dobbs Ferry, NY, 1969).
Lewis, A.D.E. and Ibbetson, D.J. (eds), *The Roman Law Tradition* (Cambridge, 1994).
Pugsley, D., *Justinian's Digest and the Compilers* (Exeter, 1995).
Rawson, B. (ed), *The Family in Ancient Rome* (London 1986; pbk. 1992).
Robinson, O.F., *The Sources of Roman Law: Problems and Methods for Ancient Historians* (London, 1997).
Robinson, O.F., Fergus, T.D. and Gordon, W.M., *European Legal History: Sources and Institutions* 2nd ed. (London, 1994).
Schulz, F., *History of Roman Legal Science* (Oxford, 1946).
———, *Principles of Roman Law* (Oxford, 1936; repr. 1967).
Stein, P., *Regulae Juris* (Edinburgh, 1966).
———, *Legal Institutions: The Development of Dispute Settlement* (London, 1984).
———, *The Character and Influence of the Roman Civil Law* (London, 1988).
———, *Römisches Recht und Europa* (Frankfurt-am-Main, 1996).
Tamm, D., *Roman Law and European Legal History* (Copenhagen, 1997).
Tellegen-Couperus, O., *A Short History of Roman Law* (London, 1993).
Thomas, P.J., Van Der Merwe, C.G. and Stoop, B.C., *Historical Foundations of South African Private Law* (Durban, 1998).
Vinogradoff, P., *Roman Law in Medieval Europe* (Oxford, 1929; repr. 1961).
Vranken, M., *Fundamentals of European Civil Law* (London, 1997).
Watson, A., *Roman Private Law around 200 B.C.* (Edinburgh, 1971).
———, *Roman Law and Comparative Law* (Athens, GA, 1991).
———, *The Spirit of Roman Law* (Athens, GA, 1995).
Wieacker, F., *A History of Private Law in Europe* trans. T. Weir (Oxford, 1995).
Wolff, H.J., *Roman Law: An Historical Introduction* (Norman, OK, 1951).
Zimmermann, R., *The Law of Obligations: Roman Foundations of the Civilian Tradition* (Cape Town, 1990).
van Zyl, D.H., *History and Principles of Roman Private Law* (Durban, 1983).

Glossary

This glossary contains a selection of Latin technical vocabulary. Italicized words indicate the translations adopted in this book. If the Latin term is not known to the reader, the English-Latin Word List immediately following this glossary may be consulted.

Abusus. *Right of disposal.* Ownership, according to one conception, is composed of the right to use (*usus*), the right to fruits (*fructus*), and the right of disposal.

Accessio. *Accession.* A mode of original acquisition of ownership, typified by the union of a principal thing and an accessory thing. The owner of the principal thing becomes owner of the whole.

Actio. *Action.* A right to bring a claim against another for legal redress. Or, the claim itself.

Actio ad exhibendum. *Action for production.* An action whose object is to force a defendant to produce a disputed thing.

Actio arbitraria. *Discretionary action.* An action in which the defendant is condemned only if he fails to comply with an order of the judge, such as an ACTIO AD EXHIBENDUM (qv).

Actio de dolo. *Action for fraud.* An action on the case permitted to one who alleged that he had suffered from the fraudulent acts of another, and to whom another action was not available. See also DOLUS; EXCEPTIO DOLI.

Actio de tigno iuncto. *Action for beams set in.* A penal action allowing a person to sue the owner of a building for double the value of stolen building materials.

Actio depensi. *Action on expenditure.* An action brought against the principal debtor by a surety who had paid the debt. The action was allowed if the principal debtor had not reimbursed the surety within six months.

Actio doli. See ACTIO DE DOLO.

Actio ex empto. *Action on purchase.* An action brought by a buyer against a seller for the seller's failure to perform his obligations under a contract of sale.

Actio ex vendito. *Action on sale.* An action brought by a seller against a buyer for the buyer's failure to perform his obligations (principally, to pay) under a contract of sale.

Actio furti concepti. See FURTUM CONCEPTUM.

Actio furti non exhibiti. See FURTUM NON EXHIBITUM.

Actio furti oblati. See FURTUM OBLATUM.

Actio furti prohibiti. See FURTUM PROHIBITUM.

Actio in factum. *Action on the case.* A non-standard action directing a judge to enter a particular judgment if he should find certain facts to be true.

Actio in personam. *Personal action.* An action founded on the existence of a relation, such as that created by a contract or delict, between two people.

Actio in rem. *Real action.* An action founded on the existence of a relation, such as ownership, between a person and a thing.

Actio noxalis. *Noxal action.* An action brought by an injured person against a master for the acts of his slave, or against a father for the acts of a son within authority, requesting alternatively the payment of reparation or the surrender of the slave or son.

Actio Publiciana. *Publician action.* A praetorian action which resembled a vindicatory action, but which directed the judge to accept as true that the duration requirements of usucapion were satisfied.

Actio utilis. *Policy action.* An action formulated on the analogy of a standard action, where the facts set out in the formula reflected the same policy which underlay the standard action.

Actus. *Drive.* A rustic praedial servitude allowing one to drive an animal or vehicle over the land of another.

Adoptio. *Adoption.* In the broad sense (encompassing ADROGATIO, qv) the act of taking a person into paternal power as a child. In the narrower sense the act whereby, in classical law, a person under the paternal power of one man came under the paternal power of another man.

Adrogatio. *Adrogation.* An act whereby a male *sui juris* (ie, not under another's paternal power) comes under another's paternal power.

Adulterium. *Adultery.* A statutory crime consisting of, strictly speaking, fornication by or with a married woman, but extending to certain other sexual acts.

Aes et libra. See PER AES ET LIBRAM.

Agnatus. *Agnate.* A person related to another via a common paterfamilias (living or dead).

Ambitus. *Electoral corruption*, or *bribery.*

Animus. *Intention.* The word denotes a state of mind whose presence or absence may be significant in a given context, eg possession, theft (ANIMUS FURANDI, qv), and ownership (ANIMUS REVERTENDI, qv).

Animus furandi. *Intention to steal.* In some sources an intention to steal is presented as a requirement of theft. In other sources it is introduced as a means of identifying an act of theft in conduct that would otherwise be regarded as innocent.

Animus revertendi. *Homing instinct.* A characteristic of particular animals, evidenced by a habit of coming back to the same place.

Annona. *Corn supply.*

Aquae ductus. *Aqueduct.* A rustic praedial servitude allowing the owner of the dominant estate to conduct water from or across the land of another.

Arra. *Down-payment.* Money or a thing given as earnest in a contract of sale.

Auctoritas tutoris. *Guardian's endorsement.* An act of consent given by a guardian to a transaction by his ward.

Bona fide. *(in) Good faith.* A state of mind, which is typified by a belief in the rightness or fairness of one's conduct, and the presence or absence of which may be significant, eg, in certain acts of possession and in the conclusion or execution of many forms of contract.

Bonorum possessio. *Estate-possession.* A praetorian scheme of succession introduced to correct the inequities of the scheme of succession under the state law.

Capitis deminutio. *Status-loss.* The term denotes any of three changes in status: loss of freedom, loss of citizenship, and removal from one's agnatic family.

Cautio. *Undertaking.* An obligation undertaken to guarantee the execution of an existing obligation or duty.

Codicilli. *Codicil.* A document which disposes of property on the death of the testator, but does not institute an heir.

Cognatus. *Cognate.* A person related to another by blood.

Commixtio. *Mixture.* A separable mingling of the property of two or more persons. If the mingling is consensual, the product is owned in common; if not, ownership is unchanged.

Commodatum. *Loan for use.* A gratuitous loan of a thing for a particular purpose.

Condictio. *Action of debt.* A personal action claiming that another should give or do something, the claim often being made in circumstances where the more usual bases for an obligation were not present; the *condictio* was therefore useful in the context of unjustified enrichment.

Confusio. *Fusion.* An inseparable mingling of the property (such as liquids) belonging to two or more persons, resulting in joint ownership of the whole.

Constitutio. *Pronouncement.* A term embracing various forms of imperial law-making.

Contrectatio. *Handling.* In many texts 'handling' is treated as an essential element of FURTUM (qv).

Corrumpere. *Spoil.* By juristic interpretation this came to describe the extent of damage to property which was actionable under the third chapter of the LEX AQUILIA (qv).

Culpa. *Fault.* The notion of fault is significant principally in the law of contract and damage to property. In some contexts *culpa* signifies a lesser degree of fault than DOLUS or CUSTODIA (qv), while in other contexts it signifies a broader range of blameworthiness, including DOLUS.

Cura. *Supervision.* In private law *cura* entailed a duty to look after the affairs of one who could not do so himself, eg, on account of some physical or mental handicap.

Curator. *Supervisor.* One who undertook the CURA (qv) of another.

Custodia. *Insurance liability.* The term signifies a particular degree of liability, higher than that of CULPA (qv), either expressly contracted or implicit in the contract (eg, COMMODATUM and LOCATIO CONDUCTIO (qv)).

Damnum emergens. *Consequential*

loss. In reckoning loss under the LEX AQUILIA (qv) account might be taken of loss to which a pecuniary value could be assigned, over and above the loss pertaining to the property concerned.

Damnum iniuria datum. *Loss wrongfully caused*. See LEX AQUILIA.

Dediticius. *Capitulated alien*. A citizen of a vanquished state who occupied a status above that of a slave but below that of a Roman citizen.

Deiectum effusumve. *Thrown or poured*. An occupier of a building was strictly liable under a quasi-delictual action for damage caused by things thrown or poured from a building (such as the contents of a chamber-pot).

Deportare. *Deport*, or *transport*. A capital form of banishment, serving as a punishment for a crime.

Depositum. *Deposit*. A gratuitous contract by conduct, under which a person looks after the property of another.

Dolus. *Fraud*, or *wicked intent*. See ACTIO DE DOLO; EXCEPTIO DOLI.

Dominium. *Ownership*. A relationship between a person and a thing often characterised by a plenary power over the thing (see ABUSUS).

Dos. *Dowry*. Goods given to the bridegroom by the bride or the bride's father pursuant to marriage.

Emancipatio. *Emancipation*. The release of a male or female from paternal power.

Emptio venditio. *Sale*. A contract by agreement in which a thing is exchanged for money.

Exceptio doli. *Defence of fraud*. An assertion by a defendant either that (1) the plaintiff had somehow acted fraudulently or in bad faith in the circumstances giving rise to the lawsuit, or (2) the lawsuit itself is unfair or not in good faith. [*Editor's note*: *exceptio* is rendered 'defence' rather than 'plea [in defence]', as in Birks and McLeod. This change was adopted to make clear that an *exceptio* is indeed a defence even when it is used as a sword.] See also ACTIO DE DOLO; DOLUS.

Exceptio non numeratae pecuniae. *Defence of money not paid*. An assertion by a defendant that he did not receive the money for which he is being sued.

Exheredatio. *Disinheriting*, or *disherison*. A testamentary act by which a person who would otherwise inherit property is prevented from doing so.

Extraneus heres. *Outside heir*. An heir such as an emancipated son who is not within the authority of the testator at the time of death.

Familiae emptor. *Property-purchaser*. The acquirer of property under a will by bronze and scales.

Fideicommissum. *Trust*. In general, a direction to a beneficiary under a will to pass on a piece of property to someone else.

Fideiussio. *Guarantee*. A form of personal security created by stipulation, creating an accessory obligation which might secure an obligation of any kind.

Fideiussor. *Guarantor*. See FIDEIUSSIO.

Filiusfamilias. *Son within authority*. A male of any age who was within the authority of a PATERFAMILIAS (qv).

Fructus. *Fruits*. (Also, *Right to fruits*. See under ABUSUS.) The produce of a thing, such as the young of animals, crops, or rents. Ordinarily the owner of the thing is the owner of the fruits.

Furtum. *Theft*. Generally, a theft was committed when one took another's property without his consent; it might occur, however, by the misuse of property entrusted to one, by taking one's own property from another with a greater right of possession, or by simply 'handling' (CONTRECTATIO, qv) the property. See also ANIMUS FURANDI.

Furtum conceptum. *Theft by receiving*. A person on whose premises stolen property was found in the presence of witnesses was liable for three times the value of the property. See also FURTUM OBLATUM.

Furtum manifestum. *Manifest theft*. A thief caught in the act, either actually so caught or deemed to be so caught, has committed manifest theft.

Furtum nec manifestum. *Non-manifest theft*. A theft committed by a thief other than a manifest thief.

Furtum non exhibitum. *Theft by retention*. A theft committed by one who refuses to hand over a stolen thing discovered in a search.

Furtum oblatum. *Theft by planting*. An action for theft by planting was available to one liable for FURTUM CONCEPTUM (qv) against the person who passed the property on to him.

Furtum prohibitum. *Theft by prohibition*. A person who refused an informal search of his premises was deemed a manifest thief.

Hereditas. *Inheritance*. In general, the collection of property, claims, and debts that a person leaves behind on death.

Hereditatis petitio. *Claim for an inheritance*. A vindicatory claim for all or part of an estate by an heir against another who holds property of the estate.

Heres. *Heir*. One who takes the place of a deceased with respect to the property, claims, and debts of the deceased.

Hypotheca. *Mortgage*. A type of real security in which the debtor retained possession and a limited right of ownership in the thing subject to the mortgage.

Impubes. *Child*.

In iure cessio. *Assignment in court*. A collusive action before a magistrate undertaken for the purpose of transferring ownership.

Inaedificatio. *Building on land*. A specific application of ACCESSIO (qv), describing the accession of a building to the land on which it stands.

Iniuria. *Contempt*. A delict characterised by an affront against a person by another, comprising both bodily injury and injury to reputation. See also DAMNUM INIURIA DATUM; INIURIA ATROX.

Iniuria atrox. *Aggravated contempt*. A contempt might be deemed aggravated if the victim was of a certain standing, if he was physically injured, or if the injury occurred in a public place. See also INIURIA.

Institutio heredis. *Appointment of the heir*. The nomination of a person as heir in a will.

Intentio. *Principal pleading*. In the formulary procedure, that part of the formula which sets forth the plaintiff's principal contention.

Interdictum. *Interdict*. An order of a magistrate which issued after proceedings of an administrative character and which directed a person to do something or not to do something.

Iter. *Passage*. A rustic praedial servitude allowing a person to walk or ride over the land of another.

Iudex. *Judge*. In civil proceedings, a private person who held no office but was appointed to adjudge a single case.

Iudex qui litem suam facit. See LITEM SUAM FACERE.

Iudicium publicum. *Criminal trial*.

Ius. *Law*, or *right*.

Ius civile. *Law of the state*. That part of the law which derived from the *Twelve Tables* and other legislation, together with interpretations of legislation and the rules developed by juristic practice.

Ius gentium. *Law of all peoples*. The law observed by all peoples, or the law developed in Rome to resolve disputes between Romans and foreigners.

Ius naturale. *Law of nature*. Used in several different senses, to denote the law dictated by natural reason, the law observed by all animals, or the law observed by all peoples.

Ius privatum. *Private law*. That part of the law concerned with relations between private persons.

Ius publicum. *Public law*. That part of the law concerned with the state, or with the relations between a person and the state.

Ius respondendi. *Right of replying*. A special distinction granted by the emperor to the legal advice of certain jurists.

Iusta causa. *Just cause*, or *legally sufficient ground*. A requirement for certain legal acts such as possession or usucapion. In general it comprises events which serve to explain or justify the act.

Lance et licio. *(with) Dish and loincloth*. A person would be adjudged a manifest thief if stolen property were found on his premises after a formal search by one wearing a loincloth and carrying a dish.

Legatarius partiarius. *Cut-in legatee*. One who receives a fraction of an estate by legacy.

Legis actio. *Action in the law*. A form of procedure that was observed from before the time of the *Twelve Tables* until the late Republic. It was characterised by its oral presentation and the strictness of its forms.

Lex. *Act*. In general, a statute enacted by an assembly of the whole Roman people, but usually including the enactments of the plebeian assembly, and sometimes including instruments issued by magistrates or the emperor.

Lex Aquilia. *Aquilian Act*. A plebiscite of the 3rd century BC allowing recovery for loss wrongfully caused (*damnum iniuria datum*), ie, for the killing of animals or slaves, and for damage to property by burning, breaking, or damaging (*urere, frangere, rumpere*). See also CORRUMPERE; DAMNUM EMERGENS.

Lex regia. *Regal Act*. An act which conferred upon the emperor the power of making laws.

Liberi. *Descendants*.

Litem suam facere. *Make the case his*

own. An obligation as though from delict, under which the misconduct of a judge who tried a case renders the judge himself liable.

Locatio conductio. *Hire*. A contract by agreement, under which a thing is hired, services are performed, or a piece of work is performed.

Lucrum cessans. *Lost gain*. The profit lost as a consequence of the loss of or damage to property actionable under the LEX AQUILIA (qv). See also DAMNUM EMERGENS.

Maiestas. *Treason*.

Mala fide. (*in*) *Bad faith*.

Mancipatio. *Mancipation*. A formal conveyance of ownership accomplished by a public act in the presence of witnesses. It was employed for the conveyance of *res mancipi*, as well as in certain other transactions such as the emancipation of a child or the making of a will. See also PER AES ET LIBRAM.

Mancipium. *Bondage*. A condition describing certain free persons who were under the power of another. The condition ordinarily subsisted temporarily, eg, during the mancipatory acts accompanying adoption or emancipation.

Mandatum. *Mandate*. A contract by agreement, under which a person undertakes to perform a service for another gratuitously.

Manumissio. *Manumission*. The release of a slave from servitude.

Manus. *Marital subordination*. A feature of a certain form of marriage, in which the wife was within the authority, similar to that of *patria postestas*, of her husband.

Merces. *Charge*. That which is paid (eg, as wages or rent) in a contract of hire.

Metus. *Duress*. A praetorian action, the *actio quod metus causa*, was extended to persons who were compelled by threats to submit to some legal transaction.

Mutuum. *Loan*. A contract by conduct, in which the borrower becomes the owner of the thing lent and returns its equivalent to the lender.

Necessarius heres. *Compulsory heir*. A slave who is freed and instituted as heir acquires the estate immediately.

Negotiorum gestio. *Uninvited intervention*. A type of obligation as though from contract, under which a person undertakes to manage another person's affairs without his authorisation.

Nomina arcaria. *Cash entries*. Entries in the account book of a Roman citizen might serve as evidence that a debt had been contracted.

Nomina transcripticia. *Account entries*. A type of contract by writing, in which entries in the account book of a Roman citizen were used to effect a novation.

Novatio. *Novation*. The extinguishment of a debt and its substitution by another.

Occupatio. *Occupation*. A type of original acquisition, under which a person becomes owner by taking possession of a thing which had no owner.

Oratio. *Speech*. A form of imperial lawmaking, in which a speech by the emperor proposing a SENATUSCONSULTUM (qv) is read in the senate.

Parens. *Ascendant.*

Parricidium. *Parricide.* The murder of one's parents or grandparents, or other close kin.

Paterfamilias. *Head of the family.* A male Roman citizen, not within the authority of another, to whom belonged the incidents of paternal power.

Patria postestas. *Paternal power*, or *family authority*. The head of the family exercised certain powers over his wife and descendants relating to, eg, the ownership of property, competence in legal transactions, consent to marriage, and noxal surrender.

Patronus. *Patron.* After manumission the master of the former slave became his patron. The patron and freedman owed certain duties to one another.

Peculium. *Personal fund.* Money or other property held by a person within authority (son or slave) for his use and disposal, but owned by the head of the family or master. See also PECULIUM CASTRENSE.

Peculium castrense. *Military fund.* Money or other property acquired by a son within authority through military service. Some freedom was available in disposing of the property notwithstanding the fact that it was owned by the head of the family. See also PECULIUM.

Per aes et libram. *By bronze and scales.* The term describes various formal acts, such as MANCIPATIO (qv) and the will by bronze and scales, involving certain spoken words, five Roman citizens as witnesses, and a scale-holder.

Perceptio. *Harvesting (of fruits).* Fruits ordinarily belonged to the owner of the thing that produced them, but certain persons, such as the usufructuary, might become owner of the fruits by harvesting. See also SEPARATIO.

Peregrinus. *Foreign*, or *peregrine*.

Permutatio. *Exchange (n).* The name given to the otherwise 'innominate' contract, under which one thing is exchanged for another.

Pignus. *Pledge.* A contract by conduct, under which a debtor transfers possession of a thing to his creditor, who holds the thing until the debt is satisfied.

Plagiarius. *Kidnapper.*

Plebiscitum. *Plebeian statute.* An enactment passed by the *concilium plebis*.

Positum aut suspensum. *Placed or hanging.* A certain obligation as though from delict imposed liability on the occupier of premises from which something was suspended to the danger of passers-by.

Postliminium. *Rehabilitation.* By this right a Roman citizen who had been enslaved by an enemy regained his freedom and some other rights on his return.

Potestas. *Authority.* See PATERFAMILIAS; PATRIA POTESTAS.

Praescriptis verbis. *(with) Special preface.* Actions with a special preface were praetorian extensions of actions on contracts by conduct, in which a preface was added at the beginning of the formula reciting the underlying events giving rise to the suit.

Pretium. *Price.* The amount a buyer was obliged to pay under a contract of sale.

Procurator. *General agent*. One who looked after the affairs of another with that person's authorisation, for example when that person was away.

Pupillus. *Ward*. See TUTELA; TUTOR.

Quasi ex contractu. *As though from contract*. Obligations based neither on an agreement between persons nor on delict, but which bore a closer relation to contractual than delictual obligations.

Quasi ex delicto. *As though from delict*. A group of offences which differed from other delicts perhaps in applying only to certain classes of persons, such as an occupier of premises or a judge.

Rapina. *Robbery*. See VI BONORUM RAPTORUM.

Recuperatores. *Assessors*. A panel of judges appointed to sit in special cases, such as those for INIURIA (qv).

Rei persequendae. *Restorative*. An action that would compensate the plaintiff for his loss, in contrast to a penal action, which would punish the defendant.

Relegare. *Relegate*, or *detain*. A non-capital form of banishment in which a person was removed from Rome and, ordinarily, his home province.

Repetundae. *Extortion*.

Res. *Thing*. One of the three principal divisions of the law. In its broadest sense it includes incorporeal things such as an inheritance and an obligation.

Res corporalis. *Corporeal thing*. A thing that exists physically, ie, that can be touched.

Res deiectae effusaeve. See DEIECTUM EFFUSUMVE.

Res derelicta. *Abandoned thing*. A thing from which the owner has parted, with the intention of not owning the thing any longer.

Res immobilis. *Immoveable thing*. Land or a building. Immoveables were distinguished from other property for certain purposes, such as usucapion, mancipation, and possessory interdicts. Property that was not land or a building and for which the distinction was significant was moveable.

Res incorporalis. *Incorporeal thing*. A thing which cannot be touched, eg, an obligation, an inheritance, or a usufruct.

Res mobilis. *Moveable thing*. See RES IMMOBILIS.

Res nullius. *Unowned thing*. A thing belonging to no one.

Res positae aut suspensae. See POSITUM AUT SUSPENSUM.

Res religiosa. *Religious thing*. The ground where a corpse was lawfully buried.

Res sacra. *Sacred thing*. A consecrated thing, such as a temple and its contents.

Res sancta. *Sanctified thing*. The gates and walls of a city.

Residuae. *Monies remaining*, or *public funds*.

Rumpere. *Damage* (*v*). See LEX AQUILIA.

Senatusconsultum. *Resolution of the senate*. A decree of the senate on a matter put before it by a magistrate or the emperor.

Separatio. *Separation (of fruits).* Fruits ordinarily belonged to the owner of the thing that produced them, but on separation of the fruits from the thing, certain persons, such as the *bona fide* possessor, might become owner instead. See also PERCEPTIO.

Servi poenae. *Penal slaves.* Persons who became slaves by condemnation.

Servi publici. *Slaves in the service of the state.* A slave who was owned by the state.

Sicarius. *Murderer.*

Societas. *Partnership.* A contract by agreement, under which several persons agree to share profits and losses.

Specificatio. *Specification.* A form of original acquisition, in which a thing belonging to a person is transformed into a new thing. In some cases this gave the maker of the new thing a claim for ownership.

Stipulatio. *Stipulation.* A contract concluded by the oral exchange of words.

Stricti iuris. *(of) Strict law.* An action in which the judge is precluded from considering matters not raised in the formula.

Suus heres. *Immediate heir.* An heir who was under the paternal power of the decedent.

Testamenti factio. *Capacity to make a will.*

Testamentum. *Will.*

Testamentum calatis comitiis. *Will before the convocation.* An archaic form of making a will before a popular assembly.

Testamentum militare. *Military will.* Soldiers were given the privilege of preparing informal wills.

Testamentum procinctum. *Will in battle-line.* An archaic form of will, prepared when battle was imminent.

Thesaurus. *Treasure.* A valuable, hidden item of moveable property, whose owner cannot be determined and which cannot be regarded as abandoned.

Traditio. *Delivery.* A form of derivative acquisition of property typified by the physical handing over of a thing, accompanied by cause.

Tutela. *Guardianship.* An institution principally for the protection of the property of minors not under paternal power and women not under paternal power.

Tutor. *Guardian.* A person who principally looked after the property of a minor not under paternal power or a woman not under paternal power.

Usucapio. *Usucapion.* A form of acquisition of ownership under which a person who possesses a thing for a period of time under certain conditions becomes owner of the thing.

Usucapio pro herede. *Usucapion as heir.* A person who possessed an item of property that formed part of an inheritance might become owner of the property under relaxed rules of usucapion.

Usus. *Right to use.* See ABUSUS.

Usus fructus. *Usufruct.* An incorporeal real right allowing the right-holder (usufructuary) to take the fruits of another's property.

Veteres. *Republican jurists.*

Vi. See VI BONORUM RAPTORUM; VIS.

Vi bonorum raptorum. *Things taken by force*. A theft of moveables accomplished by violence gave rise to the delict of *rapina*.

Via. *Way*. A rustic praedial servitude allowing the owner of the dominant estate to drive a carriage or ride over a road on the servient estate.

Vindicatio. *Vindicatory action*. A real action claiming ownership of a disputed thing.

Vis. *Force*. The use of force is relevant in several areas of both public and private law, such as possession (as in the interdict *unde vi*), robbery, contempt, and certain crimes (eg, under the *Lex Iulia de vi*).

English – Latin Word List

Abandoned thing. *Res derelicta*
Accession. *Accessio*
Account entries. *Nomina transcripticia*
Act. *Lex*
Action. *Actio*
Action for beams set in. *Actio de tigno iuncto*
Action for fraud. *Actio de dolo*
Action for production. *Actio ad exhibendum*
Action *in personam*. See PERSONAL ACTION
Action *in rem*. See REAL ACTION
Action in the law. *Legis actio*
Action of debt. *Condictio*
Action on expenditure. *Actio depensi*
Action on purchase. *Actio ex empto*
Action on sale. *Actio ex vendito*
Action on the case. *Actio in factum*
Adoption. *Adoptio*
Adrogation. *Adrogatio*
Adultery. *Adulterium*
Affront. See CONTEMPT
Agent. See GENERAL AGENT
Aggravated contempt. *Iniuria atrox*
Agnate. *Agnatus*
Appointment of the heir. *Institutio heredis*
Aqueduct. *Aquae ductus*
Aquilian Act. *Lex Aquilia*
Arbitrary action. See DISCRETIONARY ACTION
As though from contract. *Quasi ex contractu*
As though from delict. *Quasi ex delicto*
Ascendant. *Parens*
Assault. See CONTEMPT
Assessors. *Recuperatores*
Assignment in court. *In iure cessio*
Authority, guardian's. See GUARDIAN'S ENDORSEMENT
Authority. *Potestas*. See also FAMILY AUTHORITY
(in) Bad faith. *Mala fide*
Bondage. *Mancipium*
Bribery. See ELECTORAL CORRUPTION
Bronze and scales. See BY BRONZE AND SCALES
Building on land. *Inaedificatio*
Buyer's action. See ACTION ON PURCHASE
By bronze and scales. *Per aes et libram*
Capacity to make a will. *Testamenti factio*
Capitulated alien. *Dediticius*
Cash entries. *Nomina arcaria*
Cause. See JUST CAUSE
Charge. *Merces*
Child. *Impubes*
Civil law. See LAW OF THE STATE
Claim. See ACTION
Claim for an inheritance. *Hereditatis petitio*
Codicil. *Codicilli*
Cognate. *Cognatus*
Compulsory heir. *Necessarius heres*
Consequential loss. *Damnum emergens*
Constitution. See PRONOUNCEMENT
Contempt. *Iniuria*
Contempt, aggravated. See AGGRAVATED CONTEMPT
Corn supply. *Annona*
Corporeal thing. *Res corporalis*
Criminal trial. *Iudicium publicum*
Curator. See SUPERVISOR
Cut-in legatee. *Legatarius partiarius*
Damage. *Rumpere*
Damage to property. See LOSS WRONGFULLY CAUSED
Defence of fraud. *Exceptio doli*
Defence of money not paid. *Exceptio non numeratae pecuniae*
Delivery. *Traditio*
Deport. *Deportare*
Deposit. *Depositum*
Descendants. *Liberi*
Detain. See RELEGATE
(with) Dish and loincloth. *Lance et licio*
Discretionary action. *Actio arbitraria*

Disherison. See DISINHERITING
Disinheriting. *Exheredatio*
Down-payment. *Arra*
Dowry. *Dos*
Drive. *Actus*
Duress. *Metus*
Electoral corruption. *Ambitus*
Emancipation. *Emancipatio*
Endorsement, guardian's. See GUARDIAN'S ENDORSEMENT
Estate-possession. *Bonorum possessio*
Exchange. *Permutatio*
Expenditure. See ACTION ON EXPENDITURE
Extortion. *Repetundae*
Family authority. *Potestas*, or *patria postestas*
Father of the family. See HEAD OF THE FAMILY
Fault. *Culpa*
Force. *Vis*
Foreign. *Peregrinus*
Fraud. *Dolus*
Fraud, action for. See ACTION FOR FRAUD
Fraud, defence of. See DEFENCE OF FRAUD
Fusion. *Confusio*
General agent. *Procurator*
(in) Good faith. *Bona fide*
Ground. See JUST CAUSE
Guarantee. *Fideiussio*
Guarantor. *Fideiussor*
Guardian. *Tutor*
Guardian's endorsement. *Auctoritas tutoris*
Guardianship. *Tutela*
Handling. *Contrectatio*
Harvesting (of fruits). *Perceptio*
Head of the family. *Paterfamilias*
Heir. *Heres*
Heir, appointment of. See APPOINTMENT OF THE HEIR
Heir, compulsory. See COMPULSORY HEIR
Heir, immediate. See IMMEDIATE HEIR
Heir, outside. See OUTSIDE HEIR
Hire. *Locatio conductio*
Homing instinct. *Animus revertendi*
Immediate heir. *Suus heres*
Immoveable thing. *Res immobilis*
Incorporeal thing. *Res incorporalis*
Inheritance. *Hereditas*
Inheritance, claim for. See CLAIM FOR AN INHERITANCE
Institution of the heir. See APPOINTMENT OF THE HEIR
Insult. See CONTEMPT
Insurance liability. *Custodia*
Intention. *Animus*
Intention to steal. *Animus furandi*
Interdict. *Interdictum*
Judge. *Iudex*
Judge's liability. See MAKE THE CASE HIS OWN
Jurists, republican. See REPUBLICAN JURISTS
Just cause. *Iusta causa*
Kidnapper. *Plagiarius*
Law of all nations. See LAW OF ALL PEOPLES
Law of all peoples. *Ius gentium*
Law of nature. *Ius naturale*
Law of the state. *Ius civile*
Law. *Ius*
Lease. See HIRE
Legally sufficient ground. See JUST CAUSE
Loan. *Mutuum*
Loan for use. *Commodatum*
Loss of status. See STATUS-LOSS
Loss wrongfully caused. *Damnum iniuria datum*
Lost gain. *Lucrum cessans*
Make the case his own. *Litem suam facere*
Mancipation. *Mancipatio*
Mandate. *Mandatum*
Manifest theft. *Furtum manifestum*
Manumission. *Manumissio*
Marital subordination. *Manus*
Meddling. See HANDLING
Military fund. *Peculium castrense*
Military will. *Testamentum militare*
Minor. See CHILD
Mixture. *Commixtio*
Monies remaining. *Residuae*
Mortgage. *Hypotheca*
Moveable thing. *Res mobilis*
Murderer. *Sicarius*
Natural law. See LAW OF NATURE
Non-manifest theft. *Furtum nec manifestum*
Novation. *Novatio*

Noxal action. *Actio noxalis*
Occupation. *Occupatio*
Outside heir. *Extraneus heres*
Ownership. *Dominium*
Parricide. *Parricidium*
Partnership. *Societas*
Passage. *Iter*
Paternal power. *Patria potestas*
Patron. *Patronus*
Penal slaves. *Servi poenae*
Peregrine. See FOREIGN
Personal action. *Actio in personam*
Personal fund. *Peculium*
Placed or hanging. *Positum aut suspensum*
Platter and loincloth. See DISH AND LOINCLOTH
Plebeian statute. *Plebiscitum*
Pledge. *Pignus*
Policy action. *Actio utilis*
Price. *Pretium*
Principal pleading. *Intentio*
Private law. *Ius privatum*
Pronouncement. *Constitutio*
Property-purchaser. *Familiae emptor*
Public funds. See MONIES REMAINING
Public law. *Ius publicum*
Public slaves. See SLAVES IN THE SERVICE OF THE STATE
Publician action. *Actio Publiciana*
Quasi-contract. See AS THOUGH FROM CONTRACT
Quasi-delict. See AS THOUGH FROM DELICT
Real action. *Actio in rem*
Recuperators. See ASSESSORS
Regal Act. *Lex regia*
Rehabilitation. *Postliminium*
Reipersecutory. See RESTORATIVE
Relegate. *Relegare*
Religious thing. *Res religiosa*
Rent. See CHARGE
Republican jurists. *Veteres*
Resolution of the senate. *Senatusconsultum*
Restorative. *Rei persequendae*
Right. *Ius*
Right of disposal. *Abusus*
Right of replying. *Ius respondendi*
Right to fruits. *Fructus*
Right to use. *Usus*
Robbery. *Rapina*
Royal Act. See REGAL ACT
Sacred thing. *Res sacra*
Sale. *Emptio venditio*
Sanctified thing. *Res sancta*
Seller's action. See ACTION ON SALE
Separation. *Separatio*
Slaves, penal. See PENAL SLAVES
Slaves in the service of the state. *Servi publici*
(with) Special preface. *Praescriptis verbis*
Son within authority. *Filiusfamilias*
Specification. *Specificatio*
Speech. *Oratio*
Spoil. *Corrumpere*
State law. See LAW OF THE STATE
Status-loss. *Capitis deminutio*
Stipulation. *Stipulatio*
(of) Strict law. *Stricti iuris*
Supervision. *Cura*
Supervisor. *Curator*
Suretyship. *Adpromissio*
Testamentary capacity. See CAPACITY TO MAKE A WILL
Theft. *Furtum*
Theft, manifest. See MANIFEST THEFT
Theft, non-manifest. See NON-MANIFEST THEFT
Theft by planting. *Furtum oblatum*
Theft by prohibition. *Furtum prohibitum*
Theft by receiving. *Furtum conceptum*
Theft by retention. *Furtum non exhibitum*
Thing. *Res*
Things taken by force. *Vi bonorum raptorum*
Thrown or poured. *Deiectum effusumve*
Transport. See DEPORT
Treason. *Maiestas*
Treasure. *Thesaurus*
Trust. *Fideicommissum*
Tutor. See GUARDIAN
Tutorship. See GUARDIANSHIP
Undertaking. *Cautio*
Uninvited intervention. *Negotiorum gestio*
Unowned thing. *Res nullius*
Use. See RIGHT TO USE
Useful action. See POLICY ACTION
Usucapion. *Usucapio*

Usucapion as heir. *Usucapio pro herede*
Usufruct. *Usus fructus*
Vindication. See VINDICATORY ACTION
Vindicatory action. *Vindicatio*
Ward. *Pupillus*
Warrant. See ACTION
Way. *Via*
Wicked intent. See FRAUD
Will. *Testamentum*
Will, capacity to make. See CAPACITY TO MAKE A WILL
Will before the convocation. *Testamentum calatis comitiis*
Will in battle-line. *Testamentum procinctum*

Cited Passages in the *Institutes*

The table below contains references to citations in section headings and on individual pages. A reference to a section heading does not subsume a reference to a page within that section.

Table of Authorities

On references to citations in section headings, see the introductory note to 'Cited Passages in the *Institutes*', above.

Subject Index